Configuring and Installing Structured Cabling Systems

Second Edition

electrical training
IBEW · NECA ALLIANCE

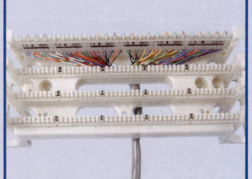

<u>Configuring and Installing Structured Cabling Systems</u> is intended to be an educational resource for the user and contains procedures commonly practiced in industry and the trade. Specific procedures vary with each task and must be performed by a qualified person. For maximum safety, always refer to specific manufacturer recommendations, insurance regulations, specific job site and plant procedures, applicable federal, state, and local regulations, and any authority having jurisdiction. The *electrical training ALLIANCE* assumes no responsibility or liability in connection with this material or its use by any individual or organization.

Acknowledgments

The *electrical training ALLIANCE* would like to thank the following individuals for their important contribution, their expertise and suggestions in the review of this textbook:

Ron Hanes, RCDD, CFOT, *Instructor*
Evansville JATC

Noel Hernberger, *Instructor*
Electrical JATC for Southern Nevada

Clint Bailey, RCDD, NICET II, *Low Voltage Systems Manager*
Commonwealth Communications

Mark Harger
Harger Lightning and Grounding

The *electrical training ALLIANCE* would also like to thank the following Training Partners for their contribution. Without their help this book would not exist:

3M
American Technical Publishers (ATP)
Channel Vision
Commscope
Corning Cable Systems
Erico, Inc.
Harger Lightning & Grounding
Ideal Networks
Leviton Manufacturing Co., Inc.
Panduit

Optional Demonstration and Hands-on Lab Equipment

This course consists of 10 instructor-led chapters along with optional hands-on exercises in the Learning Management System. In order to deliver this course in the most effective manner, obtain the items listed below "**Recommended Display Items for Class**." These items should be passed around the class at the appropriate time.

Recommended Display Items for Class	
ANSI/TIA/EIA-568, 569, 606, and J-STD-607 Standards	66-type Terminal Block
National Electrical Code	110-type Terminal Block
Backbone Cable Samples	Modular Jack Panel
Horizontal Cable Samples	Modular Cord
Outlets	110-to-110 Patch Cord
Faceplates	110-to-Modular Cord

In order to accomplish the objectives for hands-on exercises, the following equipment is required for a class of 12 or fewer students. Everything is re-usable, except for the F-type connectors. Equipment may be obtained from local distributors or from catalog sources.

Equipment Item	Quantity	Equipment Item	Quantity
66M1-50 Block	4	Wire Retention Tool for 110-type Block	8
89B Bracket for above	4	Spudger	12
100-pair 110-type Wiring Block	8	66-to-Modular Test Adapter	2
110C4 Connecting Block	50	110-to-Modular Test Adapter	2
110C5 Connecting Block	20	Wire Map Tester	1
Modular Patch Panel (Category 5e or higher)	2	Modular Outlet (Category 5e or higher)	12
1-wire Impact Tool with 66-type Blade	2	2-step Stripper for RG-6 Coaxial Cable	1
1-wire Impact Tool with 110-type Blade	6	High-quality Crimper for F-6 Connectors	1
10-wire Impact Tool for 110-type Block	2	2-piece Crimp-on F-6 Connectors	24

In addition to the equipment listed in the table above, you will need everything listed in the table titled "**Additional Materials for Hands-On Exercises**."

Additional Materials for Hands-On Exercises	
Vinyl Tape	Wood Blocks to protect table during outlet punch-down (1" × 6" × 6")
RG-6 coaxial cable	
Tie Wraps and Anchors	Plywood Backboards for 66- and 110- blocks (12" x 16")
Cross-connect wire	Continuity meter, wire map tester, or cable tester
4-pair category 5e or higher cable	25, 50, or 100-pair category 3 cable

Table of Contents

Introduction

Telecommunications has evolved to become the "Fourth Utility," after water, electricity, and gas. There are many definitions for the word *telecommunications*, but they all say essentially the same thing: Telecommunications *is the transmission and reception of any intelligence over a media*. The media may be paper, where intelligence is conveyed by writing, or electrical signals transmitted over copper cabling, radio waves transmitted through the air, or even light signals propagating through a glass fiber optic strand.

The increased demand for a reliable telecommunications infrastructure has fueled an entire industry. To keep the transmission of signals manageable and interoperable, standards for manufacturing and installation had to be introduced. The use of standards means that various devices can be connected together with the assurance that signals get to where they need to go and that vendor-specific equipment is not required to get the job done. Through the evolution of standards, "Structured Cabling Systems" were born.

Structured cabling systems were also born out of necessity. With the deregulation of the telephone companies in the 1980s, there were concerns that Ma Bell's phone equipment would not work on cabling systems they no longer had control over. The original concern was for the compatibility of phone equipment; however, as computer networks evolved and the Internet grew, structured cabling took on a whole new meaning and a whole new level of importance.

By the mid 1990s, category 5 unshielded twisted pair (UTP) was introduced; this cable allowed for data transmission rates of 100 megabits per second (Mb/s) across local area networks. The race was now on to install this cabling system everywhere. Category 5 cabling and 100 Mb/s data rates helped to fuel the telecom boom of the mid/late '90s.

As mentioned earlier, there are several types of media that can be used for the transmission of intelligence. For the purposes of a structured cabling system, the most popular and economical media types include copper cabling, fiber optics, and wireless transmission. The transmission of all intelligent signals has one thing in common: the need for providing a cabling system that will allow the transmission of signals from one active device to another. The active devices, or active electronics, convert electrical signals for transmission over the required media.

It is the purpose of the *electrical training ALLIANCE's Configuring and Installing Structured Cabling Systems* textbook to provide installers with knowledge of what a structured cabling system is, what a structured cabling system is used for, and how a structured cabling system is installed and tested. This text will focus on the installation and testing of copper cabling systems only. For further study for the installation and testing of fiber optic cabling systems, please refer to the *electrical training ALLIANCE's Reference Guide to Fiber Optics* textbook and course curriculum.

The Need for Structured Cabling Systems

Introduction

The best way to understand the need for today's structured cabling systems is to look at the history of legacy cabling systems. Many of the technologies used in these legacy cabling systems were brought forward to create current cabling systems and their installation standards. Plenty of these legacy cabling schemes still exist and are still in use. One could argue that there is really no need to study these outdated cable types, and that UTP and fiber have all but eliminated any legacy cabling. However, due in part to today's requirement for high speed and high bandwidth while still running on copper cabling, certain "legacy" type cable is still being used. Coaxial cables and shielded cables still have their place in today's cabling infrastructure.

Objectives

- Identify legacy and application-dependent type cabling systems.
- Describe multi-conductor, coaxial, twinaxial, and shielded twisted-pair cables.
- Describe N-type, BNC-type, and F-type coaxial connectors.

Chapter 1

Table of Contents

APPLICATION-DEPENDENT CABLING

Before the advent of structured cabling systems, each voice, data, or video system offered for sale to commercial users required the use of a specific cable type in order to connect the system. This type of cabling is referred to as "application-dependent cabling."

With application-dependent cabling, the designers of an equipment system dictate the specific cable type, connector type, and topology (physical layout) to be used for cabling the system. Over the years, there have been many types of cables specified for use in these systems. Unfortunately, many buildings have had multiple tenants through the years, and most tenants preferred to install new cables rather than depend on the reliability of used cabling systems. This results in layers and layers of unused cabling that ultimately create a safety hazard (fuel load from combustible cables) in the event of a fire. This issue was somewhat resolved in 2002 when the *National Electrical Code (NEC)* started to require the removal of abandoned cable.

MULTI-CONDUCTOR CABLES

A multi-conductor cable consists of a bundle of two or more conductors surrounded by a non-metallic outer jacket. **See Figure 1-1.** There may be a foil

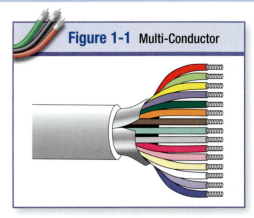

Figure 1-1 Multi-Conductor

Figure 1-1. A multi-conductor cable consists of several conductors surrounded by an outer jacket.

shield and a drain wire between the wire bundle and the outer jacket, or there may be a braided shield between the wire bundle and the outer jacket, or both. Sometimes each individual wire or pair of wires has its own foil shield, with a braided shield between the wire bundle and the outer jacket.

Multi-conductor cables have traditionally been used for wiring. Some examples are:

- Asynchronous data terminals
- Security systems
- Electronic access control systems
- Environmental controls
- Fire alarms
- Nurse call systems

DATA

Although by today's standards they are considered legacy cabling methods, there are a wide variety of data terminals still in use in commercial buildings and industrial and manufacturing facilities. There is a different type of cable for virtually every type of data terminal. It is common to see multi-conductor cables, coaxial cables, twinaxial cables, and twisted-pair cables, with or without shields, supporting the various data terminals. These cables are installed to connect data terminals to communications ports on host computers. **See Figure 1-2.**

One common type of data terminal is the asynchronous data terminal. These terminals utilize an interface described in the TIA-232-F Standard. Typically, there

For many years prior to 2002, cables accumulated above suspended ceilings.

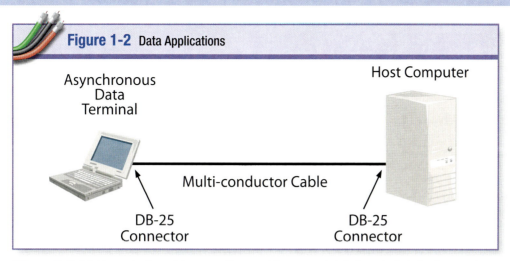

Figure 1-2 Data Applications

Figure 1-2. Multi-conductor cables are often used to connect data terminals to host computers.

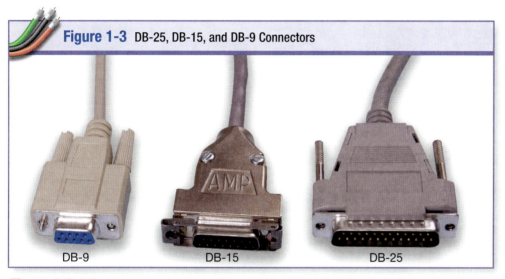

Figure 1-3 DB-25, DB-15, and DB-9 Connectors

DB-9   DB-15   DB-25

Figure 1-3. DB-style connectors are often used for serial data transmission.

is a DB-25, DB-15, or DB-9 connector located on the rear of the terminal, and a number of DB-style connectors located in the communications controller of the host computer. **See Figure 1-3.**

Anywhere from three to 12 conductors of the RS-232 interface may be required to provide suitable communications between a terminal and a host computer; this will depend on the specific implementation. These conductors are typically comprised of a multi-conductor cable. **See Figure 1-4.** It is surprising to see the number of these legacy systems still in use. The main reason that some of these systems are still being used is not the cost of replacing the

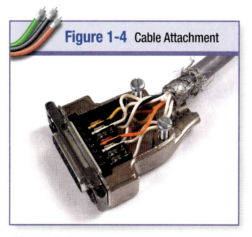

Figure 1-4 Cable Attachment

Figure 1-4. Shielded multi-conductor cable (attached to a female DB-15 connector) is commonly used in data applications.

Figure 1-5 USB Cables and Connectors

Figure 1-5. Universal serial bus (USB) cabling is used as an industry standard universal data interface.

network connection (used for longer distances within a building or campus) or via Universal Serial Bus (USB) or IEEE 1394 cable, also known as "Firewire" (used over shorter distances). The intent of USB is to eliminate all legacy connections (serial and parallel ports as well as PS/2 ports) to a computer with one universal interface. **See Figure 1-5.** USB, Firewire, and Ethernet all require the use of multi-conductor cable.

SECURITY, ELECTRONIC ACCESS CONTROL, AND LIFE SAFETY SYSTEMS

Security systems use multi-conductor cables for connecting devices, such as card readers, motion detectors, and magnetic switches, to a centralized controller. In addition, the cabling also supports control pads and speakers. **See Figure 1-6.**

Security systems and fire alarm systems have used multi-conductor cables for many years. Intrusion detection systems commonly use 4-conductor cabling for devices such as keypads, glass break

hardware but the cost and downtime of migrating to a new database program.

In general, the host computer and data terminal have been replaced with the file server and individual PCs on a shared network, or with the popular distributed computing model, which is the sharing of various servers and idle PCs across the local area network (LAN) or across the Internet. The preferred method used for the transfer of data for these systems is over the LAN on an Ethernet type of

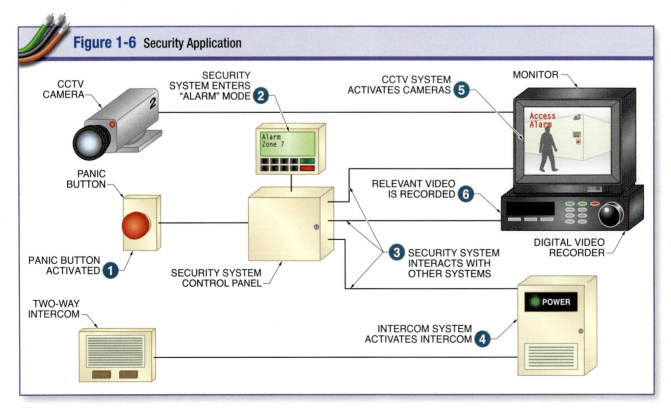

Figure 1-6 Security Application

Figure 1-6. Multi-conductor cables are often used to connect devices within a security system.

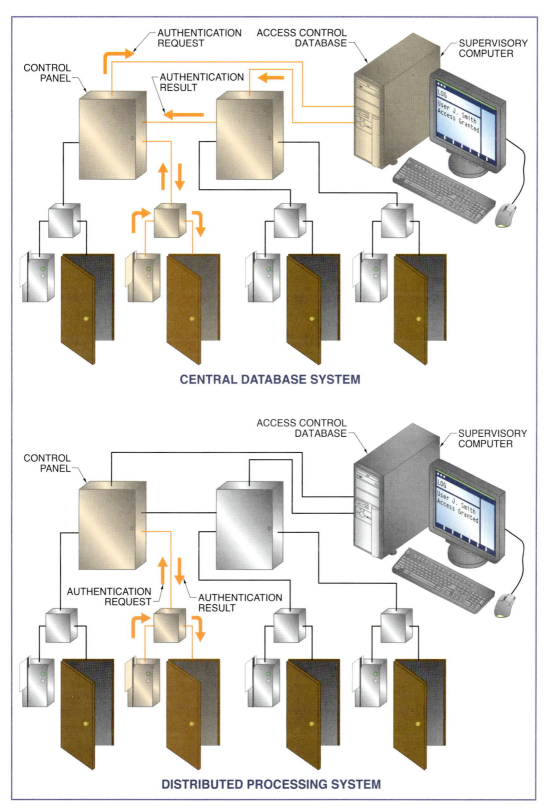

CENTRAL DATABASE SYSTEM

DISTRIBUTED PROCESSING SYSTEM

Electronic access control systems may use multi-conductor cables to send various signals and/or transfer data.

detectors, smoke detectors, motion detectors, and communications bus. Some manufacturers may require the use of a 4-conductor cable with a shield and drain wire. **See Figure 1-7.**

Access control systems traditionally have used 6-conductor cable for wiring of card readers. However, many manufacturers are now utilizing Power Over Ethernet (POE) and are allowing the use of 4-conductor cable to connect card readers. These systems use two conductors for power and two conductors for data transmission (typically RS-485 protocol). Again, the manufacturer may require shielded cable for the installation.

Intrusion detection systems also utilize 2-conductor cables for wiring door contacts, panic switches, shock sensors, and other devices. In addition, 2-conductor cable may be used for addressable loop wiring if permitted by the manufacturer. **See Figure 1-8.**

Fire alarm systems commonly use multi-conductor cables as well. Many of the fire alarm panels being installed today are addressable systems and only require 2-conductor cables. These 2-conductor cables may be required to contain a ground and a shield depending on the manufacturer's installation requirements. **See Figure 1-9.** Fire alarm panels may also utilize 4-conductor cables for remote annunciators, remote panels, or other applications. **See Figure 1-10.**

Current and future trends in environmental controls are leaning toward the use of a common network cabling scheme (structured cabling) and a common network protocol. This allows the use of "distributed control" or the direct control of HVAC devices and lighting fixtures by the use of directly connected addressable network devices. Using common open protocols such as LonWorks and BACnet removes the dependency on proprietary cabling systems for control. **See Figure 1-11.**

THE HISTORY OF STRUCTURED CABLING

There are many application-dependent (legacy) cabling schemes still in use. However, as technology progresses, most if not all of these systems will communicate over unshielded twisted-pair (UTP), shielded twisted-pair (STP), coax, or optical fiber. So it makes sense for manufacturers to standardize on a "generic" cabling system across multiple platforms.

With the breakup of the "Bells" in the mid-1980s, there was a need to "standardize" the voice cabling infrastructure because they would no longer be installing and maintaining these voice systems

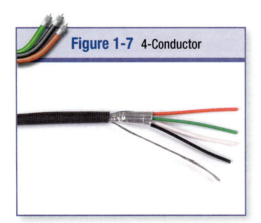

Figure 1-7. *A 4-conductor cable with shield and drain wire may be used to connect electronic keypads and/or card readers to the controller.*

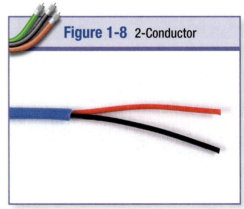

Figure 1-8. *A 2-conductor cable may be used as alarm wire, connecting devices such as magnetic contacts to the controller.*

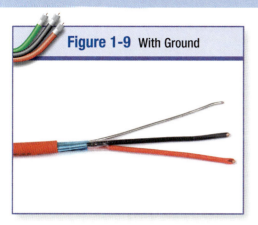

Figure 1-9. *A 2-conductor cable with ground and shield may be used in addressable fire alarm applications.*

Figure 1-10. *Fire alarm systems commonly use 2-conductor and 4-conductor cables.*

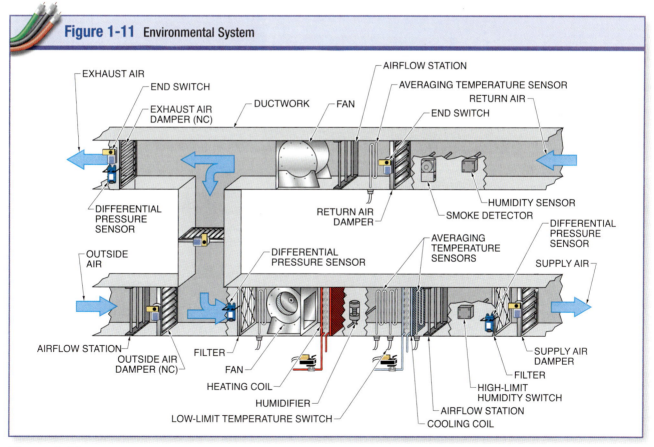

Figure 1-11. *Environmental systems employ multi-conductor cables for the interconnection of devices such as thermostats, damper controls, and fan controls.*

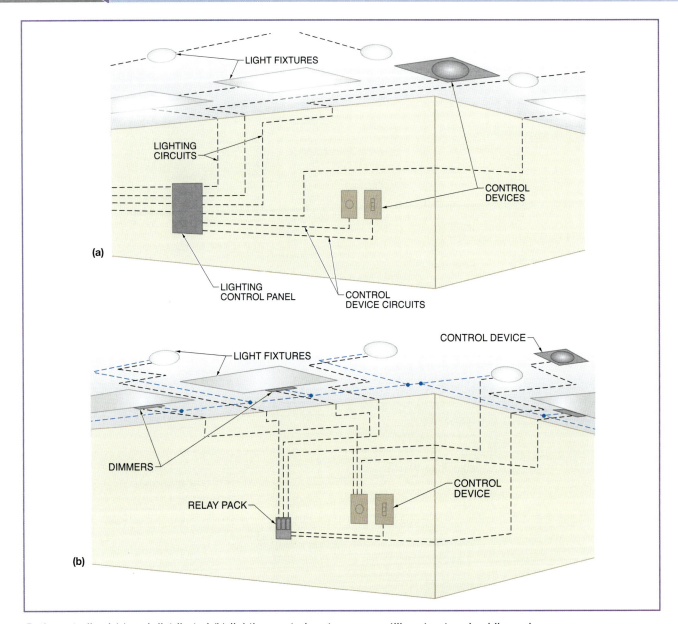

Both centralized (a) and distributed (b) lighting control systems may utilize structured cabling schemes.

for the end user. The dilemma, as the phone companies saw it, was the potential for end-users to install a variety of different cabling schemes that would not work with or may degrade the performance of the existing telephone system. This standardization would guarantee system compatibility. Standardization would also mean a cabling system that was easier to install and maintain.

At the same time, data networking systems were just beginning to show up at the customer's premise, and their perfor-

mance was dictated by modem speed and the bandwidth capability of the "plain-old-telephone" (POT) wire. Coax and optical fiber were not yet part of the "standardization." However, some of the first commercial computer networking systems were run over coax and/or shielded twisted pair. As the manufacturers of networking systems vied for prominence, end users wanted compatibility between their networking devices and did not want to change cabling infrastructure every time something new and better came along.

As computer networking progressed, three cabling technologies made the cut: the twisted pairs (UTP and STP for both voice and data); coax (for video, radio frequency (RF), and data); and optical fiber.

Coaxial Cables

A coaxial cable consists of a copper or copper-clad aluminum center conductor, surrounded by a thick insulation (a dielectric), which is in turn surrounded by an outer conductor made of solid copper or aluminum or by a metal braid. **See Figure 1-12.** A non-metallic outer insulation is optional. However, coaxial cables used indoors nearly always have a non-metallic outer insulation.

Coaxial cables may be rigid (solid outer conductor) or flexible (braided outer conductor). Flexible coaxial cables may have either solid or stranded center conductors, but they always have braided outer conductors, sometimes augmented by a foil shield over the inner insulation. **See Figure 1-13.**

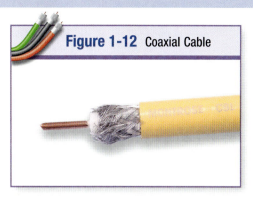

Figure 1-12 Coaxial Cable

Figure 1-12. Coaxial cable consists of a center conductor surrounded by thick insulation and a metal braid.

Coaxial cables provide better protection from electrical noise (EMI and/or RFI) and a much higher bandwidth than twisted-pair cable. In the early days of computer networking, coax was the preferred media for cabling local area networks. Today the main applications for coax are community antenna television (CATV), closed circuit television (CCTV), and other video systems as well as radio

Figure 1-13 Coaxial Cable Characteristics

Cable Type	OD (in.)	Center Conductor	Shield	Impedance
IEEE 802.3 Thin Ethernet	0.183	20 ga (19×32) stranded copper	tinned copper braid	50 ohm
IEEE 802.3 Thick Ethernet	0.405	12 ga solid copper	2 aluminum foil, 2 tinned copper braid	50 ohm
RG-59 CATV/CCTV VHF/UHF receive applications	0.193	22 ga solid copper	copper braid	75 ohm
RG-6 CATV/CCTV	0.239	21 ga copper covered steel	2 copper braids	75 ohm
RG-6 Quad CATV/CCTV	0.260	18 ga copper covered steel	2 copper braids	75 ohm
RG-11 CATV/CCTV Coaxial cable used in radio applications	0.405	14 ga solid copper	copper braid	75 ohm
RG58A/U RF transmitter applications where high power, and long runs are not required (Note: RG-58U has a solid center conductor.)	0.195	20 ga (19×32) stranded copper	tinned copper braid	50 ohm
RG 8/U RF transmitter applications	0.405	13 ga solid or stranded copper	95% copper braid	50 ohm

Figure 1-13. Coaxial cable applications vary and require specific cable types for different applications.

transmitting and receiving and other wireless and telemetry applications.

The term "RG" is an acronym for *radio guide* or *radio grade* cable. It is suggested the acronym RG be replaced with the term Series-X where the "X" designates the cable construction. For example, RG-6 would be known as Series-6 and RG-59 would be known as Series-59. This book will use the terms interchangeably since the change has not gained a foothold in the industry.

The main thing to remember when selecting coaxial cable is selecting the proper characteristic impedance of the cable to match the required application.

The characteristic impedance of series-6 (75 ohms) does not match the 50 ohm output of a radio transmitter. Selecting the series-6 cable for this application will cause severe standing waves (reflections) and a loss of power between the transmitter and antenna.

Coaxial Connectors

Coaxial cables may be terminated in soldered, screw-on, crimped, or compression connectors. **See Figure 1-14.** There are a few popular styles of coaxial connectors traditionally found in commercial buildings. The appropriate connector is chosen to match the fitting on the terminal device.

Past and present coaxial fittings include the N-type (used for thick Ethernet); the BNC-type (used in most data coaxial applications, including thin Ethernet, and in most closed circuit TV applications); and the F-type (used extensively for cable TV and broadband data applications). **See Figure 1-15.** N-type connectors may be used with larger (series-8, series-11, and 10base5) coaxial cables. N-type connectors have a center pin that must be installed over the cable's center conductor. A Baoyonet Niell - Councelman (BNC) connector is a locking type connector that was traditionally used with 10base2 thin coaxial cabling. These connectors attached to T-connectors, which in turn connected to network devices. BNC type connectors are primarily used today for connecting analog video cameras and video equipment. **See Figure 1-16.** The F-type connector is

Figure 1-14 Compression Fittings

(a) (b)

Figure 1-14. *Compression fittings (a) installed using a compression tool (b) are a more reliable termination method used today. Screw-on and crimp-type connectors are mostly relegated to the past.*

Figure 1-15 N-Type, BNC Type, and F-Type

Figure 1-15. *Coaxial connectors are chosen depending on the needs of the application.*

Figure 1-16 BNC Connector

Figure 1-16. *BNC connectors are often used to connect analog video equipment.*

Thick and Thin Ethernet LANs

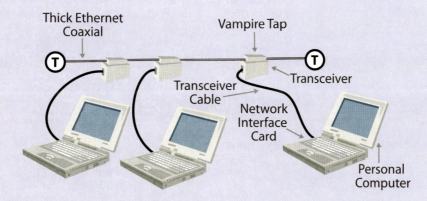

In a thick Ethernet network, thick coaxial cable, terminated at both ends with N-type connectors and 50-ohm terminating resistors, was routed past all of the locations that were connected to the network. At each location, a vampire tap was used to attach a transceiver to the cable without cutting it. A transceiver cable with a DB-15 connector on each end was used to connect the transceiver to a network interface card (NIC) located in the personal computer (PC).

This type of cabling has been replaced with UTP cabling.

Thin Ethernet cabling provided a lower-cost implementation of LAN technology compared to early Thicknet applications. With thin Ethernet, the transceiver function is built in to the NIC, thereby eliminating the need for an external transceiver, a transceiver cable, and vampire tap assembly.

Thin Ethernet cabling can be attached to a NIC using a BNC T-connector. At most locations, this type of cabling has also been replaced with UTP cabling.

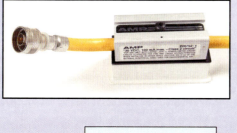

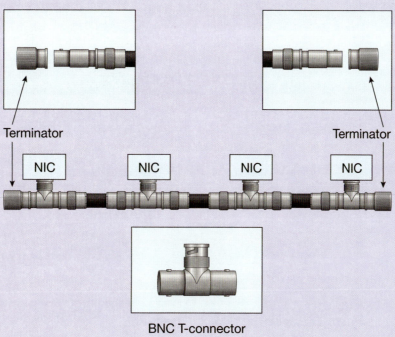

BNC T-connector

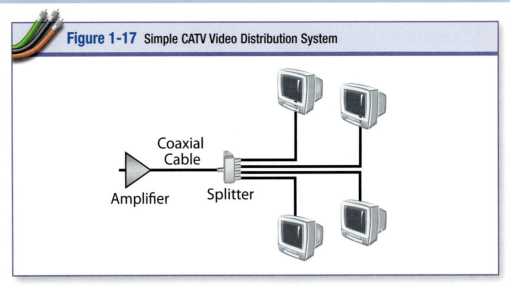

Figure 1-17 Simple CATV Video Distribution System

Figure 1-17. Because of its high bandwidth, coaxial cable is used in cable TV systems to transmit signals.

used to attach patch cables to CATV taps and other devices.

Cable TV

Cable TV wiring utilizes 75-ohm coaxial cables of various sizes. Popular types include series-6, series-6 quad, and series-11.

A simple CATV video distribution may consist of a CATV input signal, an amplifier, and a splitter to feed multiple television sets. The amplifier may or may not be needed depending on the number of television sets connected to the system and the distance to the TVs. **See Figure 1-17.**

A CATV system may also incorporate a video distribution amplifier. With this type of CATV video distribution system, the CATV signal feeds into the video amplifier and all of the CATV cabling is run in a home run fashion back to the video distribution amplifier. **See Figure 1-18.**

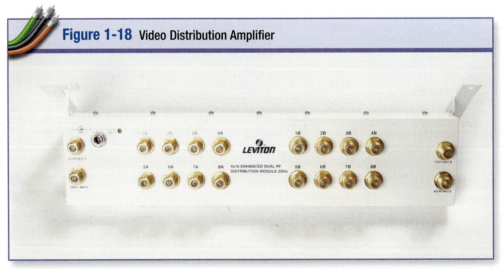

Figure 1-18 Video Distribution Amplifier

Figure 1-18. Video distribution systems use amplifiers (shown with F-type connectors) to amplify signal.

The larger the diameter of the conductor in a coaxial cable, the lower its attenuation (signal loss) will be. *Attenuation* is the decrease in magnitude of transmission signal strength between points, expressed as the ratio of output to input. It is measured in decibels (dB), usually at a specific frequency for copper or wavelength for optical fiber; the signal strength may be power or voltage.

Broadband Data

In a broadband data application, the coaxial cable is attached to a modem. In the receiving direction, the modem receives a specified band of frequencies from the coaxial cable and demodulates the signal. The demodulated signal contains data that is forwarded to a network interface card (NIC) in the PC. **See Figure 1-19.**

In the transmitting direction, data from the NIC is sent to the modem. The modem modulates a specified band of frequencies with the data and sends it out over the coaxial cable. Typically, the transmit and receive signals on the coaxial cable each occupy 12 megahertz (MHz) of bandwidth, the equivalent of two television channels. Note: In a broadband system over CATV, the distribution amplifiers and splitters must have bi-directional capabilities. **See Figure 1-20.**

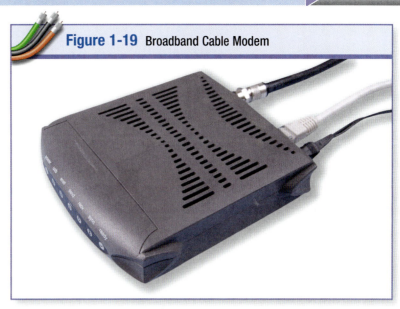

Figure 1-19 Broadband Cable Modem

Figure 1-19. The modem modulates and demodulates signals received from the NIC and the coaxial cable, respectively.

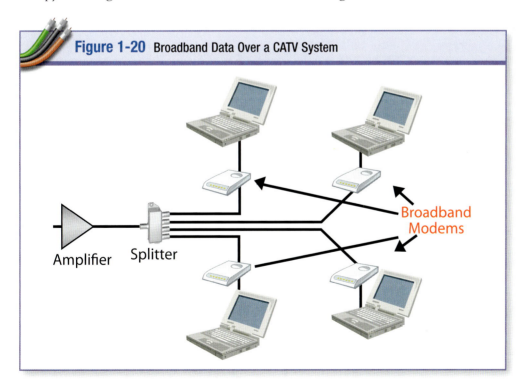

Figure 1-20 Broadband Data Over a CATV System

Broadband Modems

Amplifier   Splitter

Figure 1-20. Broadband data systems over CATV involve coaxial cables attached to a "cable" modem.

Twinaxial Cable

Twinaxial cable consists of a single, 20 AWG stranded conductor twisted pair covered by a thick plastic buffer tube. The buffer tube is covered by a braided shield, which in turn is covered with an outer insulating jacket.

The twisted pair has a characteristic impedance of 100 ohms. The twinaxial connector has two pins for terminating the signal-carrying conductors and an outer shell for terminating the cable shield.

Traditionally, twinaxial cables have been used for connecting 5250-type terminals to their controllers in an IBM AS/400 host. A single terminal or printer may be connected to a controller port, or up to seven terminals and/or printers may be daisy-chained together and connected to one controller port.

In every application, a terminator, consisting of two 50-ohm resistors and a twinaxial connector, must be installed on the unused port of the device located furthest from the controller. In the terminator, a resistor is connected between each of the conductors of the twisted pair and the shield of the cable.

Twinaxial cable consists of a single twisted pair, a plastic buffer tube, a braided shield, and an insulating jacket.

Twinaxial connectors have two pins for terminating the conductors.

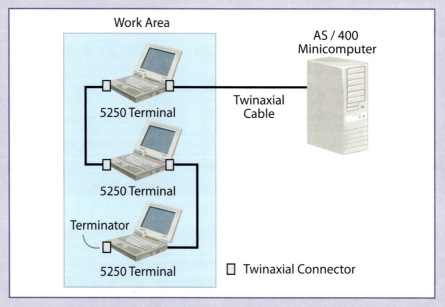

IBM AS/400 twinaxial systems, or their variations, can still be found in use today.

Shielded Twisted-Pair Cables

In the past, shielded twisted-pair (STP) cabling was a viable option for data wiring in commercial buildings, and STP cabling was endorsed by the TIA/EIA-568 (1991) and TIA/EIA-568-A (1995) Standards. However, due to tremendous improvements and wide user acceptance of unshielded twisted-pair (UTP) and fiber optic cabling systems, STP cabling fell out of favor. However, with 10Gbase-T and higher standards for data rates (40 to 100Gbase-T) and the need to solve alien crosstalk issues, using category 6A UTP and category 8 STP is becoming more popular in high-bandwidth applications.

Most shielded twisted-pair cables consist of one or more twisted pairs, with each twisted pair being covered by a shield. The shield is generally a form of aluminum foil, which may or may not be sandwiched between layers of a dielectric material such as Mylar®. Sometimes, a small (26–30 AWG) drain wire is placed inside the foil shield during the manufacture of the cable to allow a reliable point of contact for electrically grounding the shield. In other cases, a braided shield is placed around the cable core for this purpose.

Legacy Shielded Twisted-Pair Cabling.

As a replacement for coax in the LAN, IBM pioneered a twisted-pair cabling scheme that is still in use today. This was a popular shielded twisted-pair cabling system for use in the LAN, and was known as the *IBM Cabling System*. This system uses 2-pair cables for data transmission. The characteristic impedance of each twisted pair is 150 ohms. One of the pairs is colored Green/Red, while the other pair is colored Black/Orange. **See Figure 1-21.** The IBM Cabling System includes a number of STP cable types. **See Figure 1-22.**

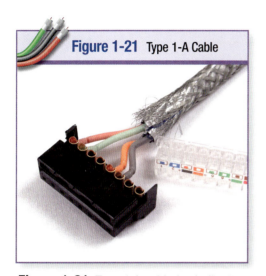

Figure 1-21 Type 1-A Cable

Figure 1-21. Type 1 A cable is similar to Type 1, but is rated for higher speeds.

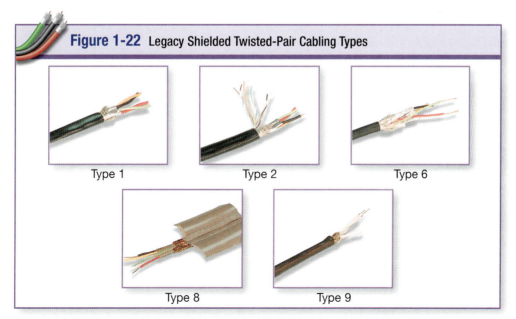

Figure 1-22 Legacy Shielded Twisted-Pair Cabling Types

Type 1   Type 2   Type 6

Type 8   Type 9

Figure 1-22. The IBM Cabling System includes a number of STP cable types.

- **Type 1** cable consists of two 22 AWG solid copper twisted pairs, each covered with a foil shield. The 2-pair bundle is covered with a braided shield.
- **Type 2** cable consists of a Type 1 cable with four 22 AWG unshielded twisted pairs between the braided shield and the outer jacket.
- **Type 6** cable consists of two 26 AWG stranded copper twisted pairs surrounded by a foil shield, a braided shield, and an outer jacket.
- **Type 8** cable consists of two parallel (non-twisted) 26 AWG solid copper pairs, each individually shielded, in a relatively flat under-carpet jacket. Type 8 cable is intended for use as an under-carpet extension of Type 1 cable.
- **Type 9** cable consists of two 26 AWG solid copper twisted pairs surrounded by a foil shield, a braided shield, and an outer jacket.

Each of these cable types is also available with an "A" suffix; for example, Type 1A. The performance of the original cables was specified at frequencies up to 20 megahertz. The performance of the newer "A" types is specified at frequencies up to 300 megahertz.

Universal Data Connector.
The legacy shielded twisted-pair cables were terminated at both ends with a 4-position universal data connector. The 4-position universal data connector terminates the insulated conductors of the two pairs, as well as the cable shield. **See Figure 1-23.** The connector is constructed in such a way that when it is not plugged into another connector, the Green/Red pair is physically looped back on the Black/Orange pair. When the connector is mated with a like connector, the loopback is opened and continuity is provided through the mated connectors for all four conductors and the shield.

At work areas, 4-position universal data connectors are mounted in faceplates. **See Figure 1-24.** In telecommunications rooms, they are mounted in distribution panels. Distribution panels are typically installed in equipment racks. However, they may be wall-mounted using special brackets. **See Figure 1-25.**

Figure 1-24 System Faceplate

Figure 1-24. *Universal data connectors are mounted in faceplates.*

Figure 1-23 Universal Data Connector

Figure 1-23. *The universal data connector is used to terminate legacy shielded twisted-pair cables.*

Figure 1-25 Cabling System Distribution Panel

Figure 1-25. In telecommunications rooms, universal data connectors are mounted on distribution panels. Courtesy of Tyco Electronics™

2-Pair STP Cabling System

The 2-pair STP cables are generally utilized between 4-position universal data connectors mounted in faceplates at work areas and 4-position universal data connectors mounted in a distribution panel in a telecommunications room. **See Figure 1-26.**

In the telecommunications room, shielded patch cords consisting of a length of Type 6 cable with a 4-position universal data connector on each end are used to connect the 2-pair STP cables to other 2-pair STP cables or to ports on electronic equipment such as token ring hubs.

Foil-shielded Cables

While the majority of 4-pair cable installations currently involve unshielded twisted-pair cables, shielded 4-pair cables are being reconsidered for high-speed data applications. Type 1A STP is capable of handling frequencies of up to 300 megahertz. Current standards for category 6 UTP rate this cable for up to 250 megahertz. IBM appears to have had it correct in 1984 when they released the original IBM Cabling System specification based on shielded cable.

The twisted pairs in a shielded 4-pair cable are the same as the pairs used in a UTP cable. However, in foil-shielded cables, the pairs are covered with a thin transparent plastic inner jacket; a drain wire is laid over the inner jacket and a foil shield is wrapped around the inner

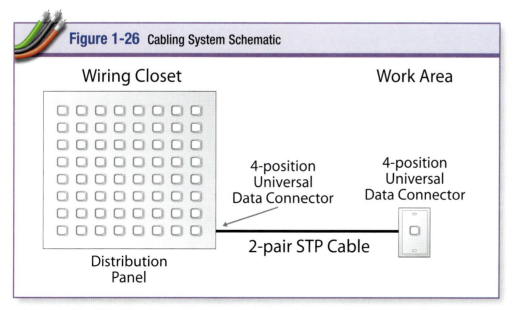

Figure 1-26 Cabling System Schematic

Wiring Closet

Work Area

Distribution Panel

4-position Universal Data Connector

4-position Universal Data Connector

2-pair STP Cable

Figure 1-26. STP cable is used to connect devices in the work area to the distribution panel by way of universal data connectors.

Old Designations	New Designations
UTP	U/UTP
FTP	F/UTP
S-FTP	SF/UTP
S-STP	S/FTP

U – Unshielded, F – Foil Shield, S – Braided Shield

U/UTP

F/UTP

S/FTP

The first letter designation indicates type of outer shield. The second letter designation indicates type of shield on each pair.

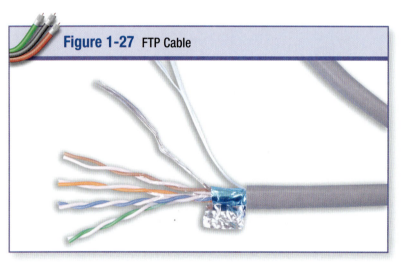

Figure 1-27 FTP Cable

Figure 1-27. *FTP cable includes an inner plastic wrapper, drain wire, foil shield, and ripcord.*

Figure 1-28 Shielded Modular Outlet

Figure 1-28. *Shielded modular outlets are used to terminate 4-pair shielded cables. Courtesy of Tyco Electronics™*

Shielded modular outlets may be assembled from several parts to ensure proper connection. Courtesy of Tyco Electronics™

jacket and the drain wire. Finally, an outer jacket is extruded around the cable. **See Figure 1-27.** Sometimes, foil-shielded twisted-pair (FTP) cables are referred to as screened twisted-pair (ScTP) cables.

Shielded Modular Outlets.

When foil-shielded 4-pair cables are used, they are terminated at work areas with shielded 8-position, 8-conductor (8P8C) modular outlets. **See Figure 1-28.** These outlets provide a method for terminating the 4-pair cables and the drain wire, and they have a metal shield for providing drain wire continuity through the outlet. Usually, a shielded modular cord is used for connecting a shielded 8P8C modular outlet to a device in a work area. In these cases, drain wire continuity is maintained through the cord, allowing the cable shield to become effectively grounded through the chassis and the green power cord ground wire of the desktop appliance.

Drain wires from 4-pair foil-shielded cables are usually grounded only at the telecommunications room. When modular patch panels are used, clips or other continuity connecting devices are provided for terminating the drain wires from individual cables and for terminating a bonding conductor, which is connected in multiple to all of the cable drain wires and to a ground conductor.

When using terminal blocks for terminating 4-pair foil-shielded cables, every fifth pair position on the block is used for grounding a cable's drain wire. Typically, the tip positions are used for terminating the drain wires. The ring positions are multiplied together and grounded. Jumper wires, a twisted pair or pairs without a jacket, are placed on the terminal block to short the tip and ring of every fifth position. These jumpers provide electrical bonds between the drain wires and ground. When wiring 110-type terminal blocks in this manner, 5-pair connecting blocks are used and up to five cables may be terminated on each row.

SUMMARY

Because computer systems traditionally installed in commercial buildings required unique application-dependent cabling systems for connecting desktop terminals to the host computers, and subsequent evolutions of the first PC LANs utilized different types of cables, building pathways were quickly filled with many cables of differing types. This created a "cabling jungle" that became unmanageable.

Structured cabling systems eliminate these "wiring jungles" by providing a uniform cabling method for all locations in a building. This uniform cabling supports voice, traditional data terminals, PC LANs, broadband data, and video services.

REVIEW QUESTIONS

1. What application has traditionally used multi-conductor cables?
   a. Electronic access control systems
   b. Environmental controls
   c. Fire alarm systems
   d. All of the above

2. Universal serial bus (USB) cable is an example of multi-conductor cable.
   a. True   b. False

3. When did the Bell companies break up?
   a. 1960s
   b. 1970s
   c. 1980s
   d. 1990s

4. Which of the following is an application for coax cabling?
   a. CATV
   b. CCTV
   c. Radio transmitting
   d. All of the above

5. ___?___ cable has a 21 ga. solid copper center conductor.
   a. RG-59
   b. RG-6
   c. RG-11
   d. RG-8U

6. Ethernet cabling provided a lower-cost implementation of LAN technology compared to early Thicknet applications.
   a. True   b. False

7. ___?___ cable consisted of two 22 AWG solid copper twisted pairs, each covered with a foil shield.
   a. Type 1
   b. Type 2
   c. Type 6
   d. Type 8

8. When did IBM originally release the IBM Cabling System specification on shielded cable?
   a. 1976
   b. 1982
   c. 1984
   d. 1991

Structured Cabling Standards

Introduction

With the breakup of the telephone companies in the mid-1980s, there was a need to ensure that cabling systems installed for voice would be installed to certain standards to ensure reliability. Up until this point the phone companies owned and installed all of the phone cables and hardware used in homes and businesses. Now that the home or business owner was responsible for the installation of equipment and cabling, the telephone companies wanted to make sure that there would be compatibility between the existing phone systems and any new cabling being installed by others.

About the same time, the computer revolution was beginning. This data was transmitted with a modem over existing telephone lines. With the growth of computers and the world wide web in 1991, data speeds were also growing. As speeds increased, the requirement for cabling infrastructure to meet this need also increased. There were many cabling schemes that were used to transmit data. Each scheme required different skill sets and different tools to install. The mid-1990s saw the increased use of unshielded twisted pair and less use of coaxial cabling. With the advent of category 5 cable, the race was on to standardize copper cable, connectors, and installation methods.

The most basic concept behind structured cabling systems is that they are designed to be "vendor independent" and to be able to meet the long-term needs of the end user. To assist in accomplishing these goals, the TIA cabling standards were established and are now the recognized industry standard. With these standards, it is possible to plan and install the necessary wiring without knowing the actual devices that will eventually be used in the building.

Objectives

- State the purpose of TIA Wiring Standards.
- Describe the function of the TIA 568, 569, 570, 606, and 607 standards.
- Describe the importance of Telecommunications Systems Bulletins.

Chapter 2

Table of Contents

STRUCTURED CABLING STANDARDS

A subcommittee of the Telecommunications Industry Association (TIA) called the TR-42 committee has published standards which have been adopted by the American National Standards Institute (ANSI). These standards serve as useful guidelines for the configuration and installation of telecommunications cabling infrastructure in commercial buildings. The purpose of these standards is to specify generic telecommunications cabling systems for commercial buildings that will support multi-product, multi-vendor environments. They also provide information that may be used for the design of telecommunications products for commercial enterprises.

In the standards, the two categories of criteria are *mandatory* and *advisory*. Mandatory requirements are designated by the word "shall." Advisory requirements are designated by the words "should," "may," or "desirable," which are used interchangeably throughout the standard. Please note that this standards document is a living document; therefore, the criteria contained in it is subject to revisions and updates as warranted by advances in building construction techniques and telecommunications technology.

ANSI/TIA-568 STANDARD

ANSI provides requirements that standards must be reviewed for revision on a five-year basis. Unfortunately, the telecommunications industry experiences change at a much more rapid pace than standards can be reviewed and updated. Because of the constant change in the industry, Telecommunications Systems Bulletins (TSBs) are issued between standard revisions. The information included in a TSB will become part of the next revised standard.

The original ANSI/TIA/EIA-568 Standard was developed and approved in 1991. Several addenda and Telecommunications Systems Bulletins were subsequently issued to address new developments within the structured cabling industry.

In 1995, the ANSI/TIA/EIA-568-A Standard was approved to incorporate all existing TSBs and addenda from the original ANSI/TIA/EIA-568 Standard.

In 2001, the ANSI/TIA-568-B Standard was approved to incorporate all TSBs and addenda attached to the 568-A version. It was also at this time that the ANSI/TIA-568 Standard was divided into three distinct parts: 568-B.1, 568-B.2, and 568-B.3. ANSI/TIA-568-B.1 addressed general requirements for a telecommunications cabling system in commercial buildings. ANSI/TIA-568-B.2 addressed balanced twisted-pair cabling and components for telecommunications cabling systems in commercial buildings. ANSI/TIA-568-B.3 addressed optical fiber cabling and components for a telecommunications cabling system in commercial buildings. It was also at this time that the EIA ceased writing standards with ANSI and TIA.

In 2009, the ANSI/TIA-568-C Standard was released as a four-part standard that included one new standard: 568-C.0, *Generic Telecommunications Cabling for Customer Premises*. The necessity to create a new Standard within the 568 document was as a result of issues that were identified with the existing structure and coverage of 568-B. Some of the issues identified were:

- Duplicate information was found within the existing 568-B Standard.
- The existing 568-B Standard had an office-oriented commercial building scope, which created gaps in standards coverage for other types of premises.
- The development process of new standards was too lengthy.

In order to address these issues, the existing information in ANSI/TIA-568-B.1 and its addenda were split between 568-C.0 and 568-C.1. With the creation of 568-C.0, a common standard was now in place to consolidate all generic structured cabling information into one document that is applicable to any type of premise. This new document provided a foundation upon which future standards could be revised. The TIA Standards are now divided into three categories: common standards, premises standards, and

Figure 2-1 Relationship Between Relevant TIA Standards

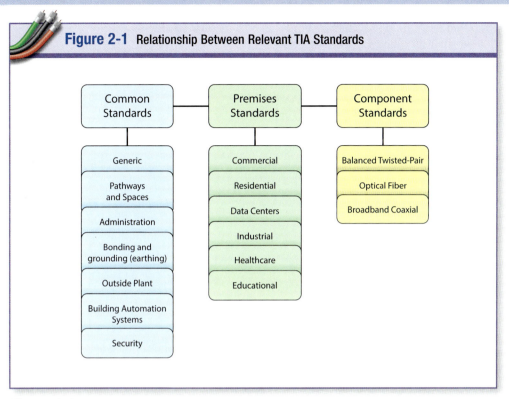

Figure 2-1. *TIA standards, including the TIA-568-D series, are divided into three categories: common, premises, and component.*

cabling and component standards. **See Figure 2-1**.

ANSI/TIA-568-C.1, *Commercial Building Telecommunications Cabling Standard,* had several new changes from the previous edition. The new 568-C.1 had new generic nomenclature found in 568-C.0. There were changes which affect inclusions of information and removal of information from the 568-C.1 document. They were:

- Inclusion of category 6 and category 6A balanced 100-ohm cabling
- Inclusion of 850 nanometer (nm) laser-optimized 50/125 micrometer (µm) multi-mode optical fiber cabling
- Inclusion of telecommunications enclosures
- Inclusion of centralized cabling in the main body of the document
- Removal of generic requirements, which have been moved to ANSI/TIA-568-C.0

- Removal of 150-ohm shielded twisted-pair cabling
- Removal of category 5 cabling
- Removal of balanced twisted-pair cabling performance testing, which has been moved to ANSI/TIA-568-C.2
- Removal of 50-ohm and 75-ohm coaxial cabling

ANSI/TIA-568-C.2, *Balanced Twisted-Pair Telecommunications Cabling and Components Standards,* had significant changes from the previous edition which included:

- Incorporation of previous TSBs, Addenda, and Interim Standards
- Definitions harmonized across all of TIA's infrastructure standards
- Performance specifications provided for category 6 and 6A balanced twisted-pair cabling
- Laboratory test measurements consolidated for all categories of cabling and components

- Field tester requirements removed from this Standard and moved to ANSI/TIA-1152

ANSI/TIA-568-C.3, *Optical Fiber Cabling Components Standard,* replaced ANSI/TIA-568-B.3, published March 2000, and ANSI/TIA-568-B.3-1, published April 2002. The ANSI/TIA-568-C.3 standard included the following significant changes from the previous edition:

- Incorporated performance specifications for 850 nanometer laser-optimized 50/125 micrometer multi-mode optical fiber cables previously found in ANSI/TIA-568-B.3-1
- Provided specifications for indoor-outdoor cable, including minimum bend radius and maximum pulling tensions
- Included array connector specifications

Significant Changes to the TIA-568 Documents

The ANSI/TIA-568 standards have been updated throughout the years as required to keep up with the ever-changing industry. Each update includes the information from TSBs between revisions while adjustments here and there have made the standards more structured and easier to understand.

One of the most noticeable changes is the naming convention used for the documents. The new naming convention changes the version "C" in the last documents to "D" in the new documents. In addition, the document version and the document number have been rearranged in the latest standards. Document 568-C.0 is now 568.0-D and so on.

TIA-568.0-D, *Generic Telecommunications Cabling for Customer Premises,* was released in September 2015 and replaces ANSI/TIA-568-C.0 and its addenda. The significant changes to the 568.0-D include:

- The annex regarding optical fiber polarity was moved to ANSI/TIA-568.3-D
- Optical fiber testing was moved to ANSI/TIA-568.3-D

- Open office cabling (i.e., consolidation points, multi-user telecommunications outlet assemblies) was moved from ANSI/TIA-568-C.1 into this standard
- Broadband coaxial cabling is now included as a recognized media in this standard
- Category 5e or higher is now required for 100-ohm balanced twisted-pair cabling
- OM3 or higher is recommended for multi-mode optical fiber cabling
- The minimum optical fiber count for multi-mode and single-mode fiber has been changed from one to two
- Formulas have been added for determining permanent link and cord lengths for balanced twisted-pair cabling
- The minimum inside bend radius for 4-pair cord cable was increased to four times the cord cable diameter
- The minimum inside bend radius for optical fiber cord cable is now specified as 25 millimeters (1 in)
- Annex E has been added to provide guidelines on shared pathways and shared sheaths for balanced twisted-pair cabling

Since the work of other committees has an impact on existing documents, it is sometimes necessary to create addendums to those existing documents. This was the case for ANSI/TIA-568.0-D. In July 2017, the TIA released TIA-568.0-D-1, *Generic Telecommunications Cabling for Customer Premises, Addendum 1: Updated References, Accommodation of New Media Types.* Significant changes in this addendum include:

- Category 8 twisted-pair cable is now an accepted media. Category 8 was specified in TIA-568-C.2-1 to 2 GHz and supports 30-meter, 2-connection channels
- TIA-568.3-D was published. Cabling transmission performance and test requirements for optical fiber cabling and guidelines for field testing of optical fiber cabling are now contained in this revision, eliminating the need to reference TIA-568-C.0 for this information.

- A new media type, OM5, was specified; therefore, OM1, OM2, and OS1 cabling is no longer supported for new installations.

TIA-568.1-D, *Commercial Building Telecommunications Infrastructure Standard,* was released in September 2015 and replaces ANSI/TIA-568-C.1 and its addenda. The significant changes to the 568.1-D include:

- Contents from Addendum 1 (pathways and spaces) and Addendum 2 (general updates) were incorporated
- Updated references
- Requirements for multi-user telecommunications outlet assemblies and consolidation points were moved to ANSI/TIA-568.0-D.
- Requirements for broadband coaxial cabling have been added
- A 2-fiber minimum count has been added for backbone cabling
- The use of optical fiber array connectors in the work area is now supported

TIA-568-C.2, *Balanced Twisted Pair Telecommunications Cabling and Components Standard,* was released in August 2009.

In November 2014, TIA released TIA-568-C.2-2, *Balanced Twisted-Pair Telecommunications Cabling and Components Standard, Addendum 2: Alternative Test Methodology for Category 6A Patch Cords.* Addendum 2 incorporated and refined the technical content of ANSI/TIA-568-C.2. Technical changes included:

- Incorporation of an alternative patch cord test head for category 6A cords
- Instructions on how to calculate limit lines for category 6A cords when using alternative test heads

In July 2016, TIA released ANSI/TIA-568-C.2-1, *Balanced Twisted-Pair Telecommunications Cabling and Components Standard, Addendum 1: Specifications for 100 Ω Category 8 Cabling.* Technical changes included:

- The addition of technical content that includes specifications for category 8 cabling

TIA-568.2-D, *Balanced Twisted-Pair Telecommunications Cabling and Components Standard,* replaced 568-C.2 and its addenda and was released in September of 2018. The significant changes include:

- Performance specs provided for category 8 shielded balanced twisted-pair cabling
- Lab test measurement methodologies updated for category 8 cabling (that may also be used for lower cabling grades)

TIA-568.3-D, *Optical Fiber Cabling and Components Standard,* was released in October 2016 and replaces ANSI/TIA-568-C.3 and its addenda. The significant changes to the 568.3-D include:

- Incorporation of optical fiber polarity content from ANSI/TIA-568.0
- Incorporation of optical fiber test measurement requirements from ANSI/TIA-568.0
- Incorporation of passive optical network component specifications
- Incorporation of polarity of cords and connectivity methods supporting parallel optical signals for transceiver interfaces and array connector patch cords and cables that exclusively employ two rows of fibers per plug
- Allows array connectivity of arbitrary row width following patterns of the illustrated 12-fiber row components
- Adds specification for wideband multi-mode fiber
- Demotes OM1, OM2, and OS 1 to *not recommended* status
- Lowers the maximum allowable OM3 and OM4 attenuation at 850 nanometer to 3.0 decibels per kilometer (dB/km)
- Raises the minimum return loss of single-mode connections and splices from 26 decibels to 35 decibels
- Accounts for the insertion loss of reference-grade test connections
- Specifies encircled flux launch cables be used for testing multi-mode connector performance at 850 nanometers

- Eliminates testing multi-mode connector performance at 1,300 nanometers
- Unifies minimum durability for all array connections to 500 mating cycles
- Adds specifications for outside plant microduct cable

In 2011, ANSI/TIA released the first edition of ANSI/TIA-568-C.4, *Broadband Coaxial Cabling and Components Standard*. It was replaced with the release of TIA-568.4-D in June of 2017. This standard specifies the requirements and recommendations for 75-ohm broadband coaxial cabling and connecting hardware to support community antenna television (CATV), satellite, and other broadband applications. The standard included such things as topology, cabling subsystems, performance, installation, and field-testing requirements. TIA-568.4-D was released in June of 2017 to replace the original 568-C.4 document. The significant changes to the 568.4-D include:

- Updates to references

Structured Cabling Architecture

Structured cabling is designed and installed according to the star topology described in the ANSI/TIA-568.0-D Standard. **See Figure 2-2.** The center of the star topology architecture is the main cross-connect (MC), also known as Distributer C. This is where the connection between the access provider (AP) and/or service provider (SP) interconnects with the building owner's structured cabling system. The MC will then either connect to an intermediate cross-connect (IC) (Distributor B), or to the horizontal cross-connect (HC), also known as Distributor A. An IC may be used to extend the cabling to an additional building in a campus-type environment or to multiple floors in a multi-story building. The backbone cabling from the MC to the IC

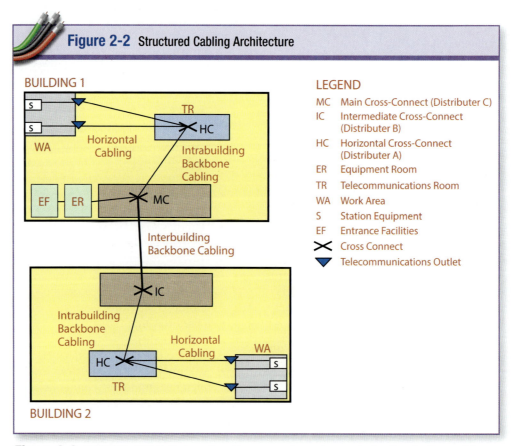

Figure 2-2 Structured Cabling Architecture

LEGEND

MC   Main Cross-Connect (Distributer C)
IC   Intermediate Cross-Connect (Distributer B)
HC   Horizontal Cross-Connect (Distributer A)
ER   Equipment Room
TR   Telecommunications Room
WA   Work Area
S   Station Equipment
EF   Entrance Facilities
✕   Cross Connect
▼   Telecommunications Outlet

Figure 2-2. ANSI/TIA-568.0-D describes the general requirements for the design and installation of structured cabling systems.

is the first level of backbone cabling. From the IC, the backbone cabling must connect to the HC. This is considered the second and final level of backbone cabling. The HC connects all work area cabling to the MC, either directly or through an IC. Backbone cabling consists of the cable and connecting hardware that provides interconnections between telecommunications rooms (TRs), equipment rooms (ERs), and entrance facilities.

Horizontal cabling connects the horizontal cross-connect to a telecommunications outlet (TO) located at a work area (WA).

By installing cross-connections at HCs, ICs, and the MC, any device on a user's desk (station equipment) may be connected to any common equipment located in a TR or in an equipment room.

STRUCTURED CABLING ELEMENTS OF A COMMERCIAL BUILDING

In order to simplify the design and specifications of a structured cabling system in a commercial building, the ANSI/TIA-568.1-D Standard lists six elements of a commercial building telecommunications cabling system. **See Figure 2-3.** They include:

1. Entrance facility
2. Equipment rooms (ER) (space typically containing Distributor C, but may contain Distributor B)
3. Telecommunications room (space typically containing Distributor A, but may contain Distributor B and Distributor C) or, in some implementations, telecommunications enclosures (space containing Distributor A)
4. Backbone cabling (Cabling Subsystem 2 and Cabling Subsystem 3)
5. Horizontal cabling (Cabling Subsystem 1)
6. Work area (space containing the equipment outlet)

Each of these elements is described and specified in a separate section of the Standard and each element is configured and designed separately.

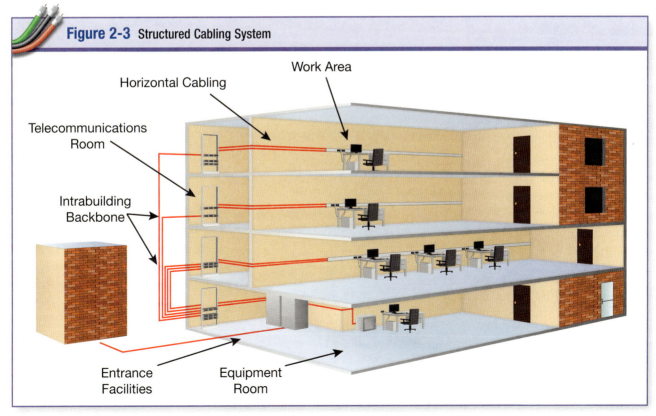

Figure 2-3 Structured Cabling System

Figure 2-3. *There are six standard elements of a basic structured cabling system.*

Work Area

The work area (WA) consists of the cabling (cords and/or adapters) that extends from the telecommunications outlet to the station equipment. It is defined in the Standards (568.1-D, Section 3.2) as *a building space where the occupants interact with telecommunications terminal equipment.* **See Figure 2-4.** The maximum length specified for work area cabling is 5 meters (16 ft).

Adapters may be necessary in the work area when the station equipment connector differs from the telecommunications outlet connector, when two or more items of station equipment must share a common telecommunications outlet and horizontal cable, or when pair transposition is necessary for compatibility. Where application-specific electrical components are required (such as impedance matching devices), they shall not be installed as part of the horizontal cabling but instead be installed externally to the telecommunications outlet.

Station equipment may include any number of devices including, but not limited to, telephone sets, data terminals, personal computers, workstations, video monitors, video cameras, and facsimile machines.

Horizontal Cabling (Cabling Subsystem 1)

Horizontal cabling includes the horizontal cable, the telecommunications outlet in the work area, and terminations and patch cords in the telecommunications room or telecommunications enclosure, and may incorporate multi-user telecommunications outlet assemblies and consolidation points. The term *horizontal* is used because these cables usually are routed horizontally in floors or ceilings and typically do not penetrate floors. Horizontal cables shall terminate on the same floor as they originate.

The standard states that a minimum of two horizontal cables shall be provided for each work area. Each 4-pair cable shall terminate in a single eight-position modular jack. The outlet for 100-ohm balanced twisted-pair cable shall meet the requirements of ANSI/TIA-568.2-D. Optical fibers at the equipment outlet shall be terminated to a duplex optical fiber outlet meeting the requirements of ANSI/TIA-568.3-D. Regardless of the

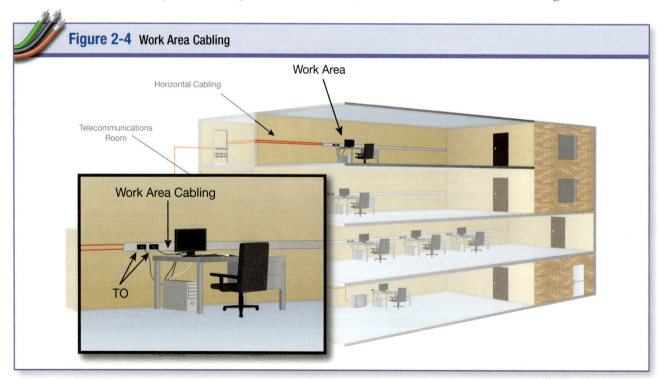

Figure 2-4 Work Area Cabling

Figure 2-4. Work area cabling includes the cabling that extends from the outlet to the equipment used by the occupants of the building.

media type, the maximum horizontal cable length shall be 90 meters (295 ft). **See Figure 2-5.**

As defined in the standards (568.2-D, Section 3.2), the horizontal cabling includes the cabling between and including the telecommunications outlet/connector and the horizontal cross-connect.

Note that the work area end of a UTP horizontal cable must terminate in one and only one outlet (jack). It is a violation of the Standard to split pairs from a 4-pair cable to multiple outlets or to install multiple appearances of any of the pairs to additional outlets.

The pathways and spaces in support of horizontal cabling shall be designed and installed in accordance with the requirements of ANSI/TIA-569-E.

Star Topology

The ANSI/TIA-568.1-D Standard requires that the horizontal cabling shall meet the star topology requirements of ANSI/TIA-568.0-D. **See Figure 2-6.** Each work area telecommunications outlet shall be connected to a horizontal cross-connect in the telecommunications room. **See Figure 2-7.**

A minimum of two horizontal cables shall be provided for each work area. According to the ANSI/TIA-568.1-D Standard, there are four types of recognized media for use in the horizontal cabling system. The recognized media are:

1. 4-pair, 100-ohm, balanced twisted-pair cabling, category 5e or higher
2. Multi-mode optical fiber cabling, 2-fiber or higher count, OM4 or higher recommended
3. Single-mode optical fiber cabling, 2-fiber or higher count
4. Broadband coaxial cabling

Horizontal Cabling Alternatives

While the majority of horizontal cabling is installed by running cables directly from a TO to a termination point in a TR, structured cabling standards allow the implementation of:

- Open office architecture (for modular furniture)
- Centralized fiber architecture (for fiber horizontals)

Open Office Architecture. Modern office designs include large areas filled with modular furniture. These spaces are frequently rearranged. These open office spaces may need to be reconfigured frequently without disturbing permanent horizontal cable runs. The ANSI/

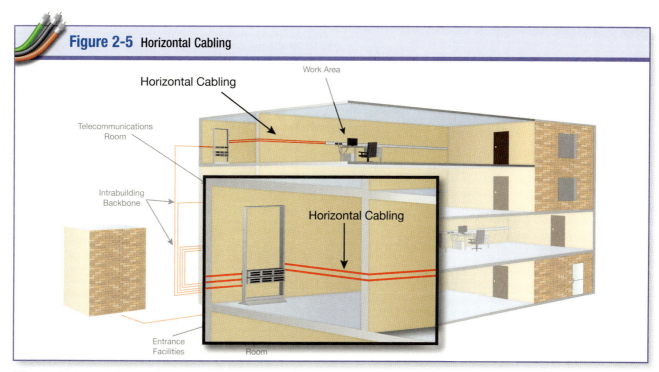

Figure 2-5 Horizontal Cabling

Figure 2-5. Horizontal cabling runs from the telecommunications room to the building outlets.

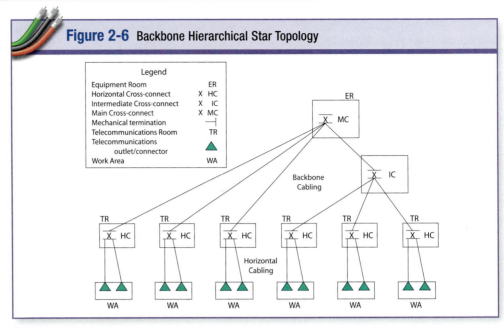

Figure 2-6 Backbone Hierarchical Star Topology

Figure 2-6. Larger networks, such as those used by large corporations or businesses, use a hierarchical star topology.

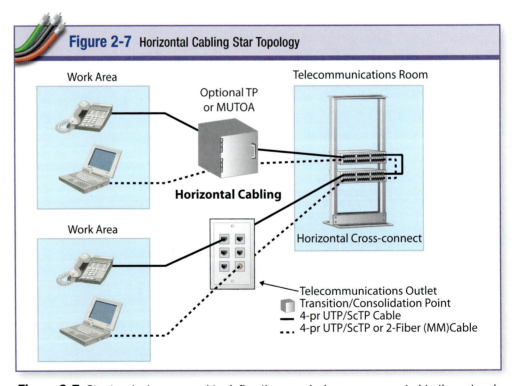

Figure 2-7 Horizontal Cabling Star Topology

Figure 2-7. Star topologies are used to define the way devices are connected to the network.

TIA-568.0-D Standard recognizes two methods for accommodating open office architecture:

- The multi-user telecommunications outlet assembly
- The consolidation point

A multi-user telecommunications outlet assembly (MUTOA) terminates multiple horizontal cables in a common location within a furniture cluster. **See Figure 2-8.** The use of MUTOAs allows horizontal cabling to remain intact when the open office plan is changed. Work area cables originating from the MUTOA are routed through modular furniture pathways and connected directly to work area equipment.

A consolidation point may be useful when reconfiguration is frequent, but not so frequent as to require the flexibility of the multi-user telecommunications outlet assembly. **See Figure 2-9.** The consolidation point is an interconnection point within the horizontal cabling. It differs from the multi-user outlet assem-

bly in that it adds an additional connection for each horizontal cable run.

An interconnection, not a cross-connection (as there is no active equipment involved), shall be used at a consolidation point. No more than one consolidation point shall be used within the same horizontal cable run. Each horizontal cable exiting the consolidation point shall be terminated on a telecommunications outlet. In addition, a transition point (going from a flat under-carpet cable to a round

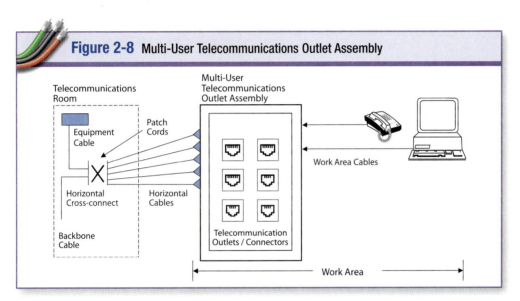

Figure 2-8. Multi-user telecommunications outlet assemblies terminate multiple cables in a common location for direct connection to station equipment.

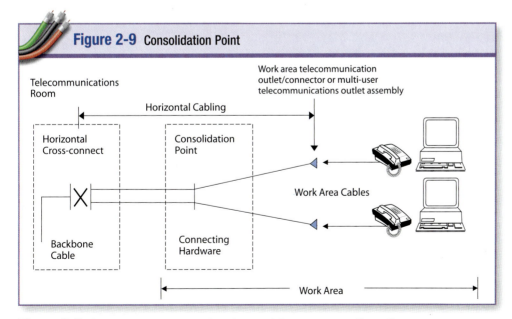

Figure 2-9. Consolidation points allow for semi-frequent reconfiguration.

cable) and consolidation point shall not be used in the same horizontal run.

For copper cabling, in order to reduce the effect of multiple connections in close proximity on near-end crosstalk (NEXT) loss and return loss, the consolidation point should be located at least 15 meters (49 ft) from the telecommunications room.

Centralized Fiber Architecture. Some users are implementing data networks with centralized electronics located in a main computer room as opposed to distributed electronics located in numerous telecommunications rooms throughout the building.

Centralized fiber architecture is an alternative to the traditional optical cross-connections located in telecommunications rooms. Centralized cabling provides connections from work areas to the centralized cross-connect by allowing the use of pull-through cables, interconnections, or splices in telecommunications rooms. **See Figure 2-10.**

Centralized cabling limits the combined length of the horizontal, intra-building backbone, and patch cords to 300 meters (984 ft). Adhering to the 300-meter limitation will ensure that the cabling system will support multi-gigabit services using centralized electronics. Centralized cabling is economical because it enables the administration of moves and changes at a centralized cross-connect location.

Backbone Cabling (Distributer B or C)

Backbone cabling is the element of the commercial building telecommunications cabling system that provides interconnections between entrance facilities (EFs), access provider (AP) spaces, service provider (SP) spaces, common equipment rooms (CERs), common telecommunications rooms (CTRs), equipment rooms (ERs), telecommunications rooms (TRs), and telecommunications enclosures (TEs). **See Figure 2-11.** The backbone cabling shall meet the requirements of ANSI/TIA-568.0-D, Cabling Distributer B and Cabling Distributer C.

Backbone cabling consists of the backbone cables, intermediate and main cross-connects (IC and MC), mechanical terminations, and patch cords or jumpers used for backbone-to-backbone cross-connection. The backbone cabling shall meet the hierarchal star topology requirements of ANSI/TIA-568.0-D.

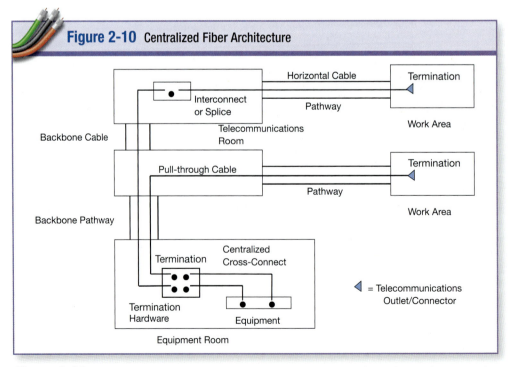

Figure 2-10 Centralized Fiber Architecture

Figure 2-10. Centralized fiber architecture provides connections from the work area to the cross-connect.

The recognized media for backbone cabling are:

- Balanced twisted-pair cabling (ANSI/TIA-568.2-D), category 5e or higher
- Multi-mode optical fiber cabling (ANSI/TIA-568.3-D), 2-fiber (or higher) fiber count; OM3 or higher recommended
- Single-mode optical fiber cabling (ANSI/TIA-568.3-D), 2-fiber (or higher) fiber count

- Broadband coaxial cabling (ANSI/TIA-568.4-D)

Backbone Cabling Maximum Lengths

Backbone cabling lengths shall be dependent upon the application and upon the specific media chosen. **See Figure 2-12.** The use of balanced twisted-pair copper cabling for backbone installations is typically limited to voice applications. The

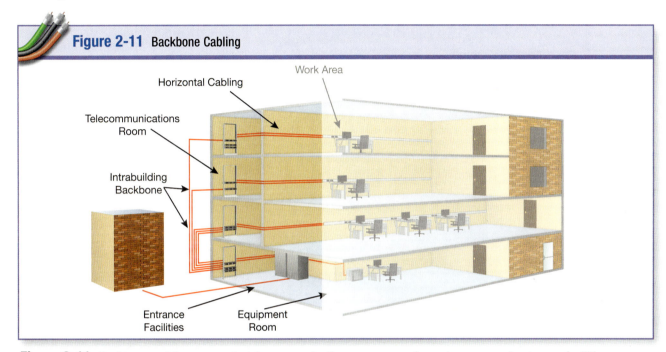

Figure 2-11 Backbone Cabling

Figure 2-11. Backbone cabling connects telecommunications rooms, equipment rooms, and entrance facilities.

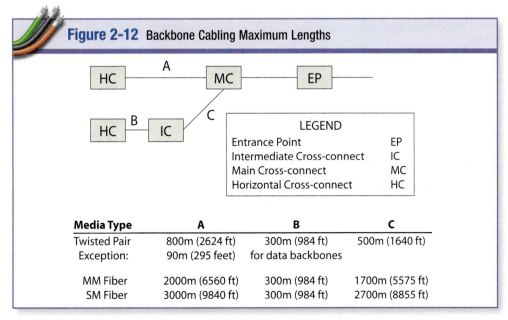

Figure 2-12 Backbone Cabling Maximum Lengths

LEGEND

Entrance Point	EP
Intermediate Cross-connect	IC
Main Cross-connect	MC
Horizontal Cross-connect	HC

Media Type	A	B	C
Twisted Pair	800m (2624 ft)	300m (984 ft)	500m (1640 ft)
Exception:	90m (295 feet)	for data backbones	
MM Fiber	2000m (6560 ft)	300m (984 ft)	1700m (5575 ft)
SM Fiber	3000m (9840 ft)	300m (984 ft)	2700m (8855 ft)

Figure 2-12. Backbone cable lengths are limited by application and media factors.

use of optical fiber cabling for backbone installations has various applications based upon performance requirements of multi-mode and single-mode optical fiber utilizing industry standard wavelengths. To maximize backbone cabling lengths, it is preferred to design the location of the MC near the center of the premises.

Maximum supportable distances for cabling can be referenced using ANSI/TIA-568.0-D, *Annex C: Application Support Information*. Table 5 lists maximum supportable distances for applications using balanced twisted-pair copper cabling. Where limited to voice applications, maximum support distances can vary from 800 meters (2,625 ft) for analog phone service to 5,000 meters (16,000 ft) for FAX, ADSL and VDSL services. Tables 6 and 7 from Annex C list maximum supportable distances and attenuation for optical fiber according to fiber type. The maximum support distances of optical fiber cabling for backbone installations vary greatly depending upon application.

For UTP cables intended as data backbones, the maximum length is the same as for a horizontal cable—90 meters (295 ft).

Telecommunications Room

A telecommunications room (TR) is an area set aside to house equipment associated with a telecommunications cabling system. The primary function of a telecommunications room is the termination of horizontal and backbone cables to compatible connecting hardware. A TR is considered floor-serving; the horizontal cables leaving the TR are terminated on the same floor where the telecommunications room is located. **See Figure 2-13.** Prior to the drafting of the ANSI/EIA/TIA-568-B.1 Standard, this room was referred to as a telecommunications closet (TC). Typically, one end of all horizontal and backbone cables terminates in a TR.

UTP cables are generally terminated on terminal blocks or on modular patch panels. Twisted-pair cross-connect wires or patch cords are used for cross-connecting

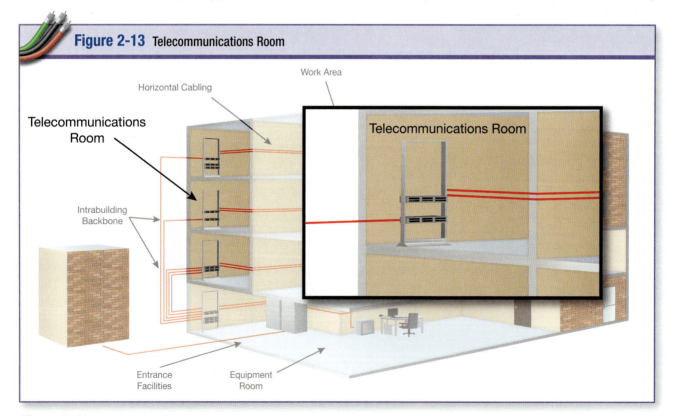

Figure 2-13 Telecommunications Room

Work Area

Horizontal Cabling

Telecommunications Room

Telecommunications Room

Intrabuilding Backbone

Entrance Facilities

Equipment Room

Figure 2-13. *Telecommunications rooms house cabling equipment and terminations for horizontal and backbone cables.*

pairs in these cables to one another or to ports on equipment located in the telecommunications room, such as a LAN switch.

Fiber optic cables are terminated in a fiber-terminating enclosure or cabinet. Fiber patch cords are used for cross-connecting fibers in these cables to one another or to ports on equipment located in the telecommunications room.

Telecommunications rooms shall be designed and provisioned according to the requirements in TIA-569-E, *Telecommunications Pathways and Spaces*.

Equipment Room

An equipment room is a centralized space for telecommunications equipment that serves the occupants of a building. **See Figure 2-14.** Equipment typically found in an equipment room may include voice switches, data switches, front-end processors, terminal controllers, LAN servers, LAN switches, and LAN routers. Equipment rooms are distinct from telecommunications rooms because of the nature of the equipment they contain (which is usually very large and serves all users in a building). However, the function of a telecommunications room may also be performed in an equipment room.

Some buildings have two equipment rooms, one for voice and one for data. Other buildings may have only one ER for all large equipment; still others could possibly have more than two ERs.

Equipment rooms shall be designed and provisioned according to the requirements in TIA-569-E. Telecommunications equipment that connects directly to main or intermediate cross-connects in the ER should do so via cables of 30 meters (98 ft) or less.

Entrance Facility

An entrance facility is the entrance to a building for both public and private network service cables, including antennae. It includes the entrance point at the building wall and continues to the en-

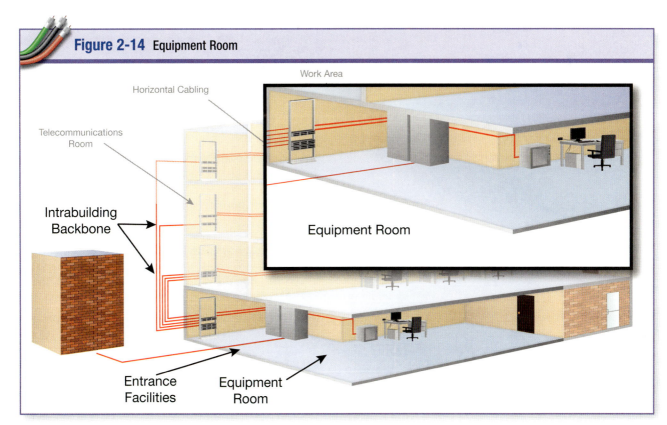

Figure 2-14 Equipment Room

Figure 2-14. An equipment room houses telecommunications equipment not found in telecommunications rooms.

trance room or space. **See Figure 2-15.** It consists of cables, connecting hardware, protection devices, and other equipment needed to connect cables entering from outdoors to cables which are suitable and approved for use indoors. An entrance facility may be in a room by itself, or it may be co-located with an equipment room or with a telecommunications room.

Entrance facilities shall be designed and provisioned according to the requirements in TIA-569-E. TIA-607-C, *Commercial Building Grounding (Earthing) and Bonding Requirements for Telecommunications,* shall also be followed.

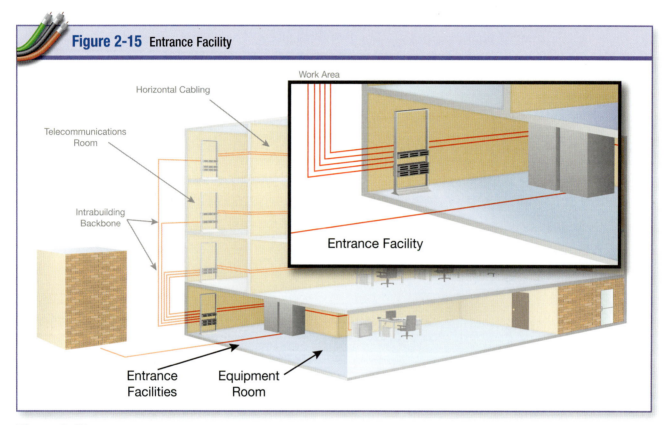

Figure 2-15 Entrance Facility

Work Area

Horizontal Cabling

Telecommunications Room

Intrabuilding Backbone

Entrance Facility

Entrance Facilities

Equipment Room

Figure 2-15. The entrance facility provides an entrance for both public and private network service cables.

SUMMARY

The purpose of the standards is to specify a generic telecommunications system for commercial and residential buildings. The standards provide the recommended practices for the design and installation of cabling systems to support a multi-product, multi-vendor environment.

By following the standards, anyone knowledgeable in the telecommunications industry will know how everything is laid out. They know where the telecommunications equipment is located, where the the backbone (distributer) cabling meets the horizontal cabling, and likely what topology is used.

Standards have no force of law like a code adopted by a state or municipality. When the standards are written into the job specifications, the installer must make sure the system is installed per the standard. It is, after all, what the customer is paying to receive.

REVIEW QUESTIONS

1. The original ANSI/TIA/EIA-568 standard was developed and approved in ___?___.
   a. 1989
   b. 1991
   c. 1994
   d. 1996

2. In 2009, the ANSI/TIA-568-C standard was released as a ___?___-part standard.
   a. 1
   b. 2
   c. 3
   d. 4

3. The ANSI/TIA-568.1-D standard lists ___?___ elements of a commercial building telecommunications cabling system.
   a. 2
   b. 4
   c. 6
   d. 8

4. What is the maximum length for work area cabling allowed in TIA-568.1-D?
   a. 1.8 m (6 ft)
   b. 3 m (10 ft)
   c. 5 m (16 ft)
   d. 7.3 m (24 ft)

5. What is the maximum horizontal cable length allowed in TIA-568.3-D?
   a. 27.4 m (90 ft)
   b. 42.7 m (140 ft)
   c. 67 m (220 ft)
   d. 90 m (295 ft)

6. Pathways and spaces in support of horizontal cabling must be designed and installed in accordance with which standard?
   a. TIA-568-C.2
   b. TIA-568.1-D
   c. TIA-569-E
   d. TIA-607-C

7. Which TIA standard requires that the horizontal cabling meet the star topology requirements of TIA-568.0-D?
   a. TIA-568-C.2
   b. TIA-568.1-D
   c. TIA-569-E
   d. TIA-607-C

8. What is the maximum supportable distance for analog phone service?
   a. 800 m (2,695 ft)
   b. 1.6 km (5,240 ft)
   c. 2.5 km (8,000 ft)
   d. 5 km (16,000 ft)

9. What is the maximum supportable distance for ADSL services?
   a. 800 m (2,695 ft)
   b. 1.6 km (5,240 ft)
   c. 2.5 km (8,000 ft)
   d. 5 km (16,000 ft)

10. The bonding and grounding of equipment rooms must adhere to which standard?
    a. TIA-568-C.2
    b. TIA-568.1-D
    c. TIA-569-E
    d. TIA-607-C

Cables and Connectors

Introduction

Electrical Workers should be familiar with the standard horizontal cable types, their associated connecting hardware, and their installation requirements and methods. The installation standards and the *National Electrical Code (NEC)* place requirements on the types of cabling and hardware to be used and how they are to be installed as part of a structured cabling system. For a cabling system to be compliant, it must meet the requirements of the standards and the applicable safety code(s).

Standard color codes for cable and connectors will be introduced; color codes for UTP, STP/ScTP, and optical fiber cable will need to be memorized.

Objectives

- Describe the differences among the various standard categories for UTP connecting hardware.
- Recognize articles/sections of the *National Electrical Code* related to telecommunications installations.
- Identify the correct color code for 4-pair UTP cables and high pair-count backbone cables.
- Relate the 4-pair UTP cable color code to the pins of an 8-pin modular jack for the T-568A and T-568B wiring configurations.
- Terminate 4-pair UTP cable onto a 66-type and a 110-type terminal block.
- Understand the basics of a fiber optic cabling system.

Chapter 3

Table of Contents

HORIZONTAL CABLING

Horizontal cabling is the portion of the telecommunications cabling system that extends from the work area telecommunications outlet to the horizontal cross-connect in the telecommunications room (TR). Horizontal cabling includes horizontal cables, telecommunications outlets in work areas, mechanical terminations, and patch cords or jumpers located in telecommunications rooms. It may also include multi-user telecommunications outlet assemblies (MUTOAs) and consolidation points (CPs).

The maximum horizontal cabling distance shall be 90 meters (295 ft), independent of media type. This is the cable length from the mechanical termination of the media at the horizontal cross-connect in the telecommunications room to the telecommunications outlet in the work area.

The length of the cross-connect jumpers and patch cords in the cross-connect facilities, including horizontal cross-connects, jumpers, and patch cords that connect horizontal cabling with equipment or backbone cabling, should not exceed five meters (16 ft) in length. **See Figure 3-1.**

The ANSI/TIA-568.1-D Standard requires horizontal cabling to be imple-mented in a star topology. **See Figure 3-2.** Each work area telecommunications outlet shall be connected to a horizontal cross-connect in a telecommunications room. Each work area should be served by a telecommunications room located on the same floor.

Horizontal cabling shall contain no more than one transition point or consolidation point between the horizontal cross-connect and the telecommunications outlet.

A *transition point* is a location where flat undercarpet cable connects to round horizontal cable. A *consolidation point* is a location where horizontal cables extending from a TR are interconnected to horizontal cables extending into modular furniture pathways.

Bridged taps and splices shall not be permitted as part of the copper horizontal cabling. (A *bridged tap* is multiple appearances of a cable pair at several termination points. A *splice* is a permanent joining of conductors, generally from separate sheaths, using mechanical means.)

Horizontal Cables

Four types of media are recognized by ANSI/TIA-568.1-D for use in the horizontal cabling system. They are:

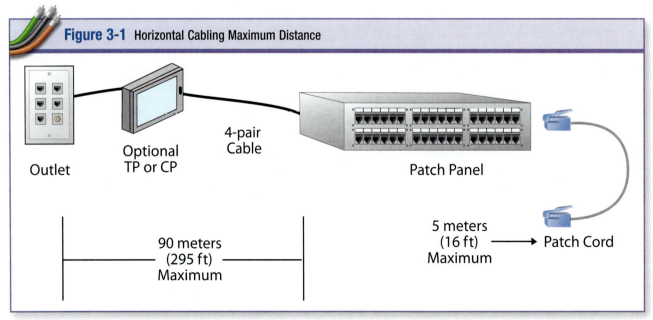

Figure 3-1 Horizontal Cabling Maximum Distance

Outlet

Optional
TP or CP

4-pair
Cable

Patch Panel

90 meters
(295 ft)
Maximum

5 meters
(16 ft) ⟶ Patch Cord
Maximum

Figure 3-1. *The maximum distance is 90 meters for horizontal cabling and five meters for jumpers, patch cords, and cross-connects.*

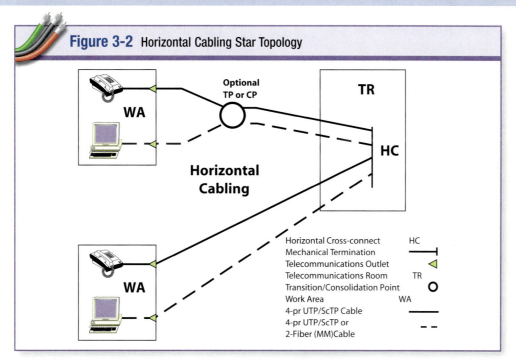

Figure 3-2 Horizontal Cabling Star Topology

Horizontal Cross-connect   HC
Mechanical Termination
Telecommunications Outlet
Telecommunications Room   TR
Transition/Consolidation Point   O
Work Area   WA
4-pr UTP/ScTP Cable
4-pr UTP/ScTP or
2-Fiber (MM)Cable

Figure 3-2. *The ANSI/TIA-568.1-D Standard requires horizontal cabling to be implemented in a star topology.*

1. 4-pair balanced twisted-pair cabling, category 5e, 6, 6A, or 8. To support a wide range of applications, category 6A cabling may be required. Category 8 performance can only be achieved with a maximum of two connections and a maximum length of 30 meters (98 ft).
2. Multimode optical fiber cabling, 2-fiber or higher count, OM3 or higher recommended
3. Single-mode optical fiber cabling, 2-fiber or higher count
4. Broadband coaxial cabling

Recognized cables, associated connecting hardware, jumpers, patch cords, equipment cords, and work area cords shall meet all applicable requirements specified in ANSI/TIA-568.2-D, 568.3-D, and 568.4-D.

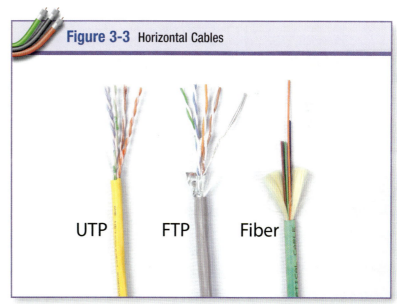

Figure 3-3 Horizontal Cables

UTP   FTP   Fiber

Figure 3-3. *Three types of recognized horizontal cables are unshielded twisted-pair (UTP), foil twisted-pair (FTP), and fiber.*

Shielded Cable Types

ScTP, *screened twisted pair,* utilizes a metallic screen or braid over the group of four pairs to provide alien crosstalk (AXT) and electromagnetic interference (EMI) shielding. **FTP,** *foil twisted pair,* is similar but uses a thin metallic foil as opposed to a braid. The difference between these two cabling types is that the screen shield is more durable and easier to terminate onto a shielded jack, while the foil shield provides better coverage (no holes) and is effective at higher frequencies but can easily tear if mishandled. **See Figure 3-3.** Another type of shielded cable is *pairs in metal foil* (PiMF). PiMF cables have a foil shield for each pair to virtually eliminate internal crosstalk and usually include an overall foil or screen to provide additional EMI immunity. **See Figure 3-4.**

Figure 3-4 Cable Designations

Old Designation	New Designation	Description
UTP	U/UTP	Unshielded Twisted Pair
FTP	F/UTP	Foil over UTP
S-FTP	SF/UTP	Screen and Foil over UTP
S-STP	S/FTP	Screen over Foil Shielded Pairs

Figure 3-4. *The mechanical construction of category cable shielding is identified with an acronym.*

Figure 3-5 Modular Outlets and Faceplate

Figure 3-5. *Per ANSI/TIA-568.1-D, two telecommunications outlets must be supplied for each work area.*

TELECOMMUNICATIONS OUTLETS

A minimum of two telecommunications outlets shall be provided for each individual work area. One telecommunications outlet may be associated with voice and the other with data. Consideration should be given to installing additional outlets based on present and projected needs. **See Figure 3-5.**

Each 4-pair cable at the equipment outlet shall be terminated in an 8-position modular jack. The telecommunications outlet/connector for 100-ohm balanced twisted-pair cable shall meet the requirements of ANSI/TIA-568.2-D.

Optical fiber cables at the equipment outlet shall be terminated to a duplex optical fiber outlet/connector meeting the requirements of ANSI/TIA-568.3-D.

Modular Outlets

Modular outlets are defined by the number of positions available in the outlet and the number of pins provided in the outlet. For example, the outlets installed in most residences in the past were 6-position, 4-conductor (6P4C) outlets. This means that the molded plastic housing was capable of holding six pins. However, in this example, only the center four positions are equipped with pins.

Today, the 8-position, 8-conductor (8P8C) modular outlet is the standard worldwide for installation in commercial buildings. It is compatible with 4-, 6-, and 8-position plugs and accommodates 1-, 2-, 3-, and 4-pair circuits.

These 8P8C outlets are frequently referred to as RJ45-style jacks. This language comes from a particular telephone company application where, historically, an 8P8C jack (called RJ45) was installed at the network interface.

At work areas, 8P8C modular outlets are mounted in faceplates or in surface-mount boxes. **See Figure 3-6.** When used for terminating cables in telecommunications rooms and equipment rooms, the outlets may be mounted in plastic or metal patch panels. These patch panels are typically installed in equipment racks. However, they may be wall-mounted by using special brackets.

Figure 3-6 8P8C Modular Outlet

Figure 3-6. *The 8-position, 8-conductor (8P8C) modular outlet is commonly installed in commercial buildings.*

Modular Outlet Configurations

There are various methods for terminating twisted pairs on 8-pin modular outlets. The ANSI/TIA-568.2-D Standard specifies two configurations. These two configurations are known as T-568A and T-568B. Both termination configurations are acceptable for voice and high-speed data.

Another method for terminating a 4-pair cable on a modular outlet is to use the Federal Communications Commission's Universal Service Order Code (USOC) RJ61X configuration. This configuration is acceptable for voice and low-speed data. It is not acceptable for high-speed data because of excessive pair-to-pair crosstalk caused by the nesting of the pairs at the outlet. The USOC configuration is no longer recognized by the ANSI/TIA Standards. **See Figure** 3-7.

When terminating 4-pair UTP cables, follow the manufacturer's instructions and remove only as much cable jacket as necessary.

Pair twists must be maintained to within 13 millimeters (0.5 in) of the point of termination for category 5 and higher cables, and within 76 millimeters (3 in) for category 3 cables. However, it is a best practice to always maintain the pair twist to the point of termination.

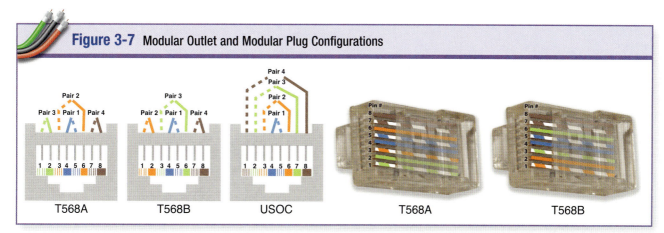

Figure 3-7 Modular Outlet and Modular Plug Configurations

Figure 3-7. *The three most common modular outlet configurations are T-568A and T-568B (for 8-pin modular outlets) and USOC (for terminating 4-pair cables).*

Color Codes

The conductor insulation of each pair in a 4-pair cable is coded with a unique color combination. **See Figure 3-8.** Some cables utilize solid colors for conductor and pair identification. In this case, each tip conductor has white insulation while every ring conductor has solid colored insulation.

Some cables utilize a white insulation with colored stripes at regular intervals on every tip conductor and a colored insulation with white stripes at regular intervals on every ring conductor. **See Figure 3-9.**

NATIONAL ELECTRICAL CODE

The *National Electrical Code (NEC)* covers installations of electrical conductors and equipment within or on public and private buildings or other structures.

The *NEC* has specific articles that apply to telecommunications cables and systems. **See Figure 3-10.** Within the telecommunications articles, acceptable cable types for the systems are identified. **See Figure 3-11.**

Definitions

- **Approved** - Acceptable to the authority having jurisdiction. This authority is usually represented by an electrical inspector, building inspector, or fire marshal (*NEC* Article 100).

- **Listed** - Equipment or materials included in a list published by an organization acceptable to the authority having jurisdiction and concerned with product evaluation, that maintains periodic inspection of production of listed equipment or materials, and whose listing states either that the equipment or material meets appropriate standards or has been tested and has been found suitable for use in a specified manner (*NEC* Article 100). UL is the most prominent organization that lists electrical equipment and cables.

- **Listed Products** - Those products that have been evaluated with respect to reasonably foreseeable hazards to life and property, and where such hazards have been safeguarded to an acceptable degree.

- **Riser** - Cables installed in vertical runs and penetrating more than one floor, or cables installed in vertical runs in a shaft.

- **Plenum** - A compartment or chamber to which one or more air ducts are connected and that forms part of the air distribution system.

Figure 3-8 UTP Color Code

Pair No.	Conductor	Solid Color	Band Striped	T568A	T568B	USOC
		4 Pair UTP Cable Color Code		8-pin Modular Jack Pin out		
Pair 1	Tip	White	White/Blue	5	5	5
Pair 1	Ring	Blue	Blue/White	4	4	4
Pair 2	Tip	White	White/Orange	3	1	3
Pair 2	Ring	Orange	Orange/White	6	2	6
Pair 3	Tip	White	White/Green	1	3	2
Pair 3	Ring	Green	Green/White	2	6	7
Pair 4	Tip	White	White/Brown	7	7	1
Pair 4	Ring	Brown	Brown/White	8	8	8

Figure 3-8. Color codes are used to identify conductor pairs within a UTP cable.

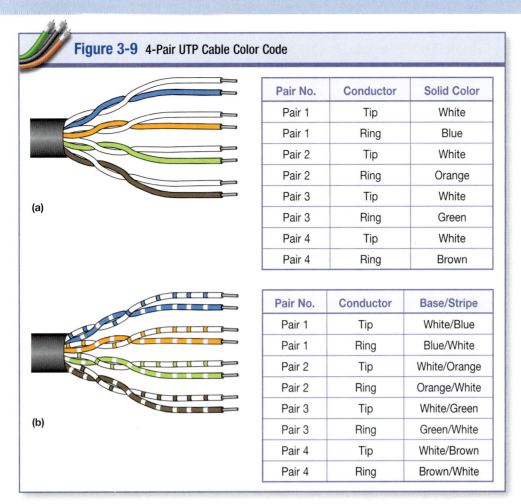

Figure 3-9   4-Pair UTP Cable Color Code

Pair No.	Conductor	Solid Color
Pair 1	Tip	White
Pair 1	Ring	Blue
Pair 2	Tip	White
Pair 2	Ring	Orange
Pair 3	Tip	White
Pair 3	Ring	Green
Pair 4	Tip	White
Pair 4	Ring	Brown

Pair No.	Conductor	Base/Stripe
Pair 1	Tip	White/Blue
Pair 1	Ring	Blue/White
Pair 2	Tip	White/Orange
Pair 2	Ring	Orange/White
Pair 3	Tip	White/Green
Pair 3	Ring	Green/White
Pair 4	Tip	White/Brown
Pair 4	Ring	Brown/White

Figure 3-9. *Each pair in a four-pair cable is color-coded, using either solid colors (a) or stripes (b).*

Figure 3-10 *National Electrical Code* Limited Energy Articles

Article	Title	Typical Applications
725	Class 1, Class 2, Class 3 Remote-Control, Signaling, and Power-limited Circuits	Data Transmission Industrial Controls Door Bells
760	Fire Alarm Systems	Fire Alarm
770	Optical Fiber Cables	Fiber Networks
800	General Requirements for Communications Systems	Telephone
805	Communications Circuits	Communications Circuits and Equipment
810	Radio and Television Equipment	Antenna Systems for Radio and TV
820	Community Antenna Television and Radio Distribution Systems	Cable TV
830	Network-Powered Broadband Communications Systems	Cable TV
840	Premises-Powered Broadband Communications Systems	Multimedia

Figure 3-10. *The* NEC *has nine articles that are used for the limited energy/telecommunications industry.*

Figure 3-11 Cable Types

Article	725	760	770	800	820	830
PLENUM	CL3P, CL2P	FPLP	OFNP, OFCP	CMP	CATVP	BLP
RISER	CL3R, CL2R	FPLR	OFNR, OFCR	CMR	CATVR	BMR
GENERAL PURPOSE	CL3, CL2	FPL	OFN, OFC	CM, CMG	CATV	BM
RESIDENTIAL	CL3X, CL2X			CMX	CATVX	BLX

Figure 3-11. The NEC establishes cable type requirements for installations.

Requirements for Cables Run in Risers and Plenums

Cables run in a riser (vertical run penetrating more than one floor) must be:

- Type CMR, CATVR, CL3R, CL2R, FPLR, OFNR, OFCR, BMR, CMP, CATVP, CL3P, FPLP, OFNP, OFCP, or BLP
- Type CM, CATV, CL3, CL2, FPL, OFN, OFC, BM, CMX, CATVX, CL3X, CL2X, or BLX in metal conduit

Note: The penetration of one floor only is allowed for Type CM , CATV, CL3, CL2, FPL, OFN, OFC, BM, or MP cable (by itself, not in conduit), but not in the same floor penetration as Type CMR, CATVR, CL3R, CL2R, FPLR, OFCR, BMR, CMP, CATVP, CL3P, FPLP, OFNP, OFCP, or BLP cables.

Cables run in a plenum must be:

- Type CMP, CATVP, CL3P, FPLP, OFNP, OFCP, or BLP
- Type CMR, CATVR, CL3R, CL2R, FPLR, OFNR, OFCR, CM, CATV, CL3, CL2, FPL, OFN, OFC, BLX, CMX, CATVX, CL3X, or CL2X in metal conduit

See *NEC* Sections 800.154 and 800.179.

Note: Cable substitutions must be acceptable to the authority having jurisdiction (AHJ). **See Figure 3-12.**

Interior-Rated Cables

Interior-rated cables come in four different varieties:

- Plenum
- Riser
- General Purpose
- Residential

All cables placed in locations covered by the *NEC* shall be listed. (See *NEC* 800.113(A)).

However, the *NEC* (805.48, Informational Note No. 2) permits up to 50 feet of non-listed outside plant cable entering the building to run exposed inside the building, if a primary protector is not required. However, if primary protection is required, and it is practicable to place the protector closer than 50 feet to the entrance point, the outside plant cable may not be extended up to 50 feet.

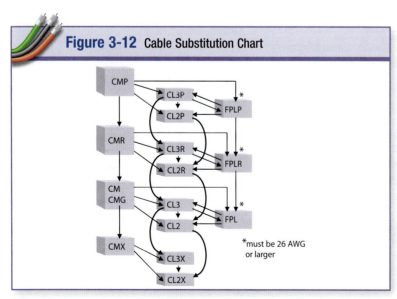

Figure 3-12 Cable Substitution Chart

*must be 26 AWG or larger

Figure 3-12. Cable substitutions must be acceptable to the authority having jurisdiction (AHJ).

The *NEC* is primarily concerned with the flammability hazards associated with cables, specifically the initiation of a fire by electrical circuits and the carrying of fire by cables. Communications circuits carry such low amounts of energy that they are incapable of initiating a fire in a fire-resistant cable. Therefore, the primary flammability hazard of communications cables is the potential for fire spread. In addition to flame spread, the requirements for cables in air-handling systems (plenums) take smoke production into consideration. There are no combustion toxicology requirements in the *NEC*.

Point of Entrance

Within a building, the point at which the wire or cable emerges from an external wall, from a concrete floor slab, or from a rigid metal conduit or an intermediate metal conduit must be grounded to an electrode in accordance with *NEC* 800.100(B). An exception to this would be the use of an all-dielectric fiber optic cable to provide services; that is, one that has no metal or metallic shielding or metal strength members.

Terminal Blocks

At work areas, 8-pin modular outlets are used for terminating 4-pair UTP cables, and they are sometimes used on the telecommunications-room end of the cable. Some manufacturers use stamped metal patch panels that the modular outlets can snap into. Terminating modular patch panels uses the same techniques that are used to terminate work area modular outlets.

When modular outlets are not used for terminations in the telecommunications room, the 4-pair cables are usually terminated on terminal blocks or 110-type patch panels. One such terminal block is the 66-type terminal block, which has been used for many years in voice installations. **See Figure 3-13.**

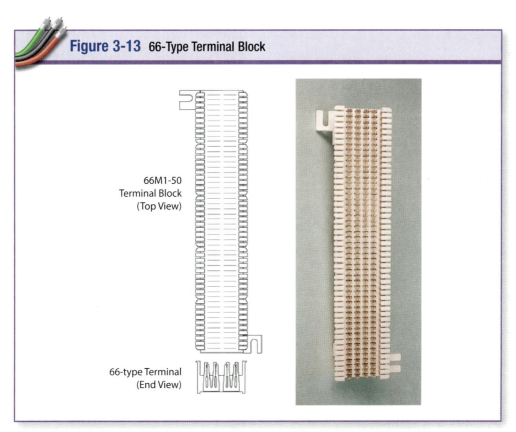

Figure 3-13 66-Type Terminal Block

66M1-50
Terminal Block
(Top View)

66-type Terminal
(End View)

Figure 3-13. 66-type terminal blocks are commonly used for voice installations.

66-Type Terminal Block

Most 66-type terminal blocks are rated category 3. However, some newer models with category 5e and 6 ratings are available. These 66-type terminal blocks come in many different configurations. While available in sizes from 4 to 50 pairs, the most popular model is the 66M1-50. This model allows up to 25 pairs to be terminated on the left side of the block and up to 25 pairs to be terminated on the right side.

Each vertical column, left and right, contains fifty 66-type connectors. The outside positions of the 66-type connectors are used for terminating wire pairs. The inside positions are used for terminating cross-connect wires.

When terminating 4-pair UTP cables on a 66M1-50 terminal block, the first

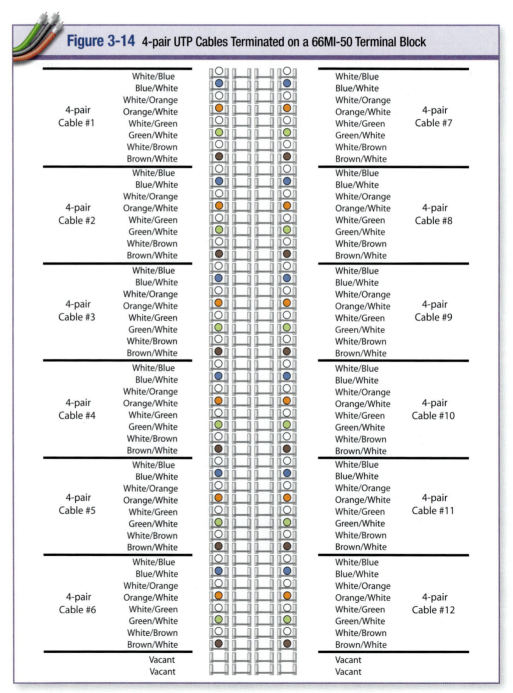

Figure 3-14 4-pair UTP Cables Terminated on a 66MI-50 Terminal Block

Figure 3-14. *Up to twelve 4-pair cables may be terminated on a 66M1-50.*

six cables should occupy positions 1 - 48 on the left outside column. The next six cables should occupy positions 1 - 48 on the right outside column. Positions 49 and 50 of both columns should remain vacant. **See Figure 3-14.**

110-Type Terminal Block

The 110-type terminal block is a typical style used for terminating horizontal and backbone cables. These blocks may have a category 5e or higher performance rating, depending upon the particular model and manufacturer.

110-type terminal blocks are available in 25-, 50-, 100-, and 300-pair sizes. The plastic wiring block is used for terminating wire pairs. Each horizontal row is

capable of terminating up to 25 pairs. **See Figure 3-15.**

110-Type Connecting Block

After the wire pairs have been seated and trimmed, 3-, 4-, or 5-pair 110-type connecting blocks must be installed on top of the wires. These connecting blocks contain metal connectors which provide electrical continuity between the wires and the positions on top of the connecting block which are used to access the wires from the terminated cable with cross-connect wires or patch cords. Typically, 4-pair connecting blocks are used when terminating 4-pair UTP horizontal cables on 110-type terminal blocks. **See Figure 3-16.**

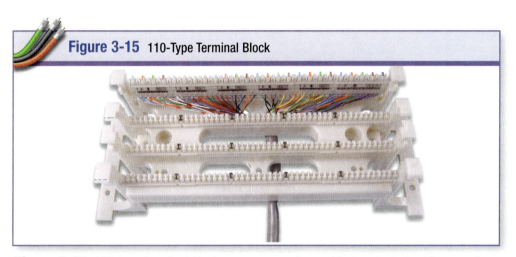

Figure 3-15   110-Type Terminal Block

Figure 3-15. *110-type terminal blocks are commonly used to terminate backbone and horizontal cables.*

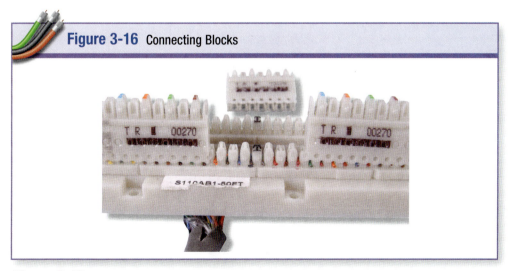

Figure 3-16   Connecting Blocks

Figure 3-16. *Connecting blocks are installed on top of the wires to provide continuity between the wires and the terminated cable.*

4-Pair UTP Cables Terminated On a 110-Type Terminal Block

A 110-type terminal block can accommodate up to six cables per row. Each cable has a specific color code that must be followed, starting with cable # 1. Note that the very last pair slot (far right) is vacant.

Cable #1	Cable #2	Cable #3	Cable #4	Cable #5	Cable #6
bl or gr br	bl or gr br	bl or gr br	bl or gr br	bl or gr br	bl or gr br vac

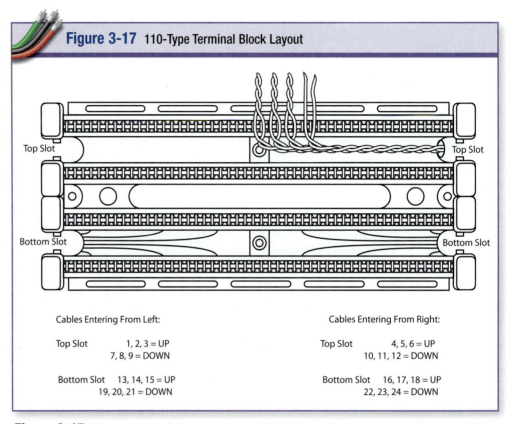

Figure 3-17 110-Type Terminal Block Layout

Top Slot

Top Slot

Bottom Slot

Bottom Slot

Cables Entering From Left:

| Top Slot | 1, 2, 3 = UP |
| | 7, 8, 9 = DOWN |

| Bottom Slot | 13, 14, 15 = UP |
| | 19, 20, 21 = DOWN |

Cables Entering From Right:

| Top Slot | 4, 5, 6 = UP |
| | 10, 11, 12 = DOWN |

| Bottom Slot | 16, 17, 18 = UP |
| | 22, 23, 24 = DOWN |

Figure 3-17. When using a 110-type terminal block, manufacturer's instructions must be carefully followed to ensure proper termination.

When terminating 4-pair UTP cables on 110-type terminal blocks, be sure to follow the cable manufacturer's and the hardware manufacturer's recommendations for the stripping of and the laying of the 4-pair cables. **See Figure 3-17.** The 4-pair cable bundles are routed through the wiring channel to the appropriate position on the terminal block and terminated.

Position 25 of each row is left vacant. Some manufacturers allow the cable jacket to be cut off at the edge of the block. Others require the jacket to be left on as close as possible to the point of conductor termination.

The conductors are seated and trimmed, using a 110 blade on an impact tool. Over each 4-pair group, 4-pair 110-type connecting blocks are installed, using a 5-pair impact tool. **See Figure 3-18.**

As an option, the rightmost 4-pair group and the vacant position may be covered with a 5-pair 110-type connecting block instead of a 4-pair version to offer a neater appearance. **See Figure 3-19.**

Patch Panels

Patch panels may be used for terminating the telecommunications-room end of the 4-pair UTP cables. Available with category 3 and higher performance, patch panels are very popular for terminating cables that are intended for data use. Because most data equipment utilizes

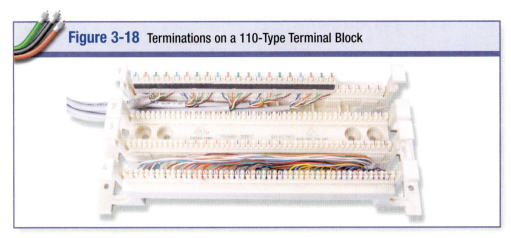

Figure 3-18 Terminations on a 110-Type Terminal Block

Figure 3-18. Both 4-pair and 25-pair cables can terminate on a 110-type terminal block; both cable types use 110-type connecting blocks.

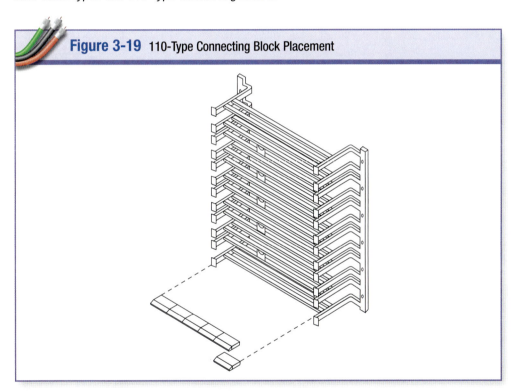

Figure 3-19 110-Type Connecting Block Placement

Figure 3-19. The rightmost 4-pair and the vacant position are sometimes covered together with the use of a 5-pair, rather than a 4-pair, connecting block.

8-position modular jacks, when a patch panel is used for terminating 4-pair cables, a simple 4-pair modular patch cord of the proper category is all that is required for patching. **See Figure 3-20.**

Many patch panels utilize printed wiring boards. Typically, a group of six 8P8C modular jacks are mounted on the front of the board and six 110-type connecting blocks are mounted on the rear of the board. A number of these boards are then mounted on the rear of a punched metal panel to create 24-, 48-, 72-, or 96-port patch panels.

When terminating a 4-pair UTP cable on a 110-to-modular patch panel, be sure to follow the cable and hardware manufacturer's instructions when it comes to the removal of the cable jacket and the placement of the 4-pair UTP cable within the patch panel. **See Figure 3-21.**

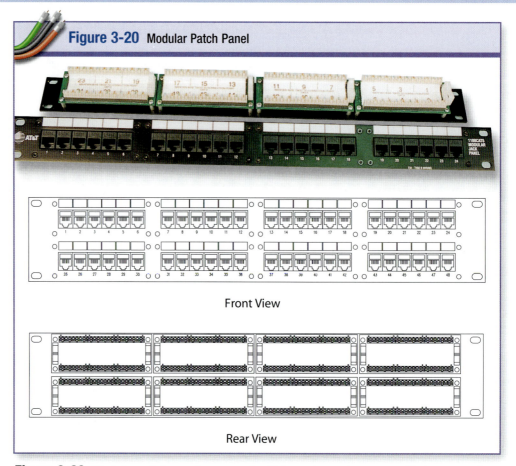

Figure 3-20 Modular Patch Panel

Front View

Rear View

Figure 3-20. *Modular patch panels are used within equipment rooms and telecommunications rooms to terminate cables.*

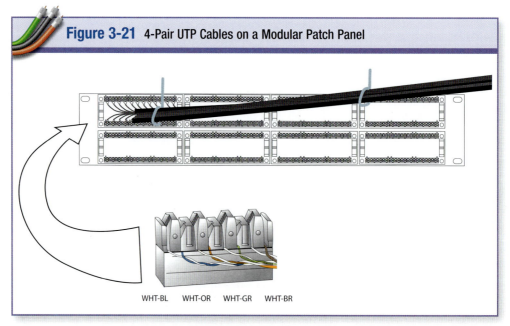

Figure 3-21 4-Pair UTP Cables on a Modular Patch Panel

WHT-BL   WHT-OR   WHT-GR   WHT-BR

Figure 3-21. *Both the cable and patch panel manufacturer's instructions must be followed to ensure proper placement of the cable.*

To terminate the pairs on a 110-type patch panel, the pairs are positioned in the grooves of the connecting block. Then the wires are seated and trimmed, using a single-wire impact tool. **See Figure 3-22.**

Patch Cords

In high-speed data environments, connections among terminal blocks, patch panels, and equipment ports are usually accomplished with patch cords. These cords are available in many combinations including 110-to-110, 110-to-modular, and modular-to-modular. **See Figure 3-23.**

UTP patch cords are available in 1-pair, 2-pair, 3-pair, and 4-pair varieties. UTP patch cables are made from stranded conductors, rather than solid conductors, for flexibility. In order to be rated category 3, 5, 5e, 6, or 6A, they must meet or exceed the transmission requirements for 4-pair horizontal cables of the same category. Because it uses stranded conductors, the patch cord attenuation is allowed to exceed the cable attenuation requirement by 20%.

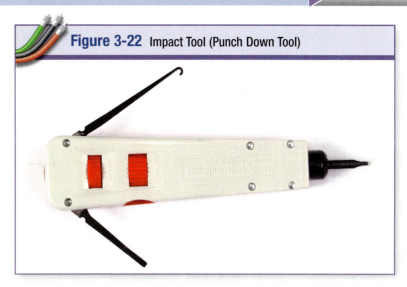

Figure 3-22 Impact Tool (Punch Down Tool)

Figure 3-22. Impact tools are used to seat and trim the wires once they have been placed on a terminal block.

UTP Cross-Connect Wire

UTP cross-connect wire is generally used for cross-connecting between terminal blocks for voice circuits and low-speed (10 megabits per second (Mb/s) or lower) data circuits. UTP cross-connect wire is generally not recommended for cross-connecting higher-speed data signals be-

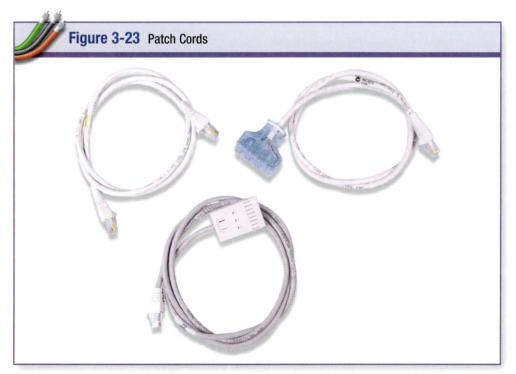

Figure 3-23 Patch Cords

Figure 3-23. Patch cords are used to connect terminal blocks, patch panels, and equipment ports.

Figure 3-24 UTP Cross-Connect Wire

Figure 3-24. Cross-connect wire is used for cross-connecting in low-speed environments.

cause it is not possible to control the degree of crosstalk among jumper wires. UTP cross-connect wire is available in 1-pair, 2-pair, 3-pair, and 4-pair varieties. **See Figure 3-24.**

Backbone Cabling

Four types of media are recognized by ANSI/TIA-568.1-D for use in the backbone cabling system. They are:

1. Balanced twisted-pair cabling
2. Multi-mode optical fiber cabling, 2-fiber or higher count; OM4 or higher recomended

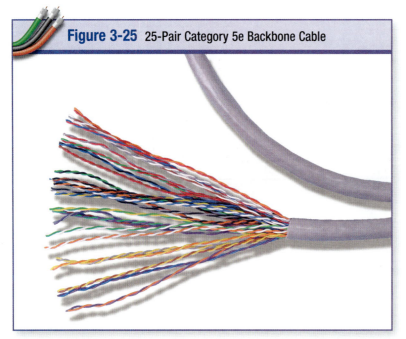

Figure 3-25 25-Pair Category 5e Backbone Cable

Figure 3-25. Each pair in a 25-pair cable has its own unique color code.

3. Single-mode optical fiber cabling, 2-fiber or higher count
4. Broadband coaxial cabling

The specific performance characteristics for the recognized cables, associated connecting hardware, and cross-connect jumpers and patch cords are described in the ANSI/TIA-568.2-D (UTP) and ANSI/TIA-568.3-D (fiber) Standards.

The maximum supportable distances are application- and media-dependent.

To minimize cabling distances, it is often advantageous to locate the main cross-connect near the center of a site. Installations that exceed these distance limits may be divided into areas, each of which can be supported by backbone cabling within the scope of the standard.

The length of UTP backbone cabling that supports data applications shall be limited to a total of 90 meters (295 ft).

In the main and intermediate cross-connects, jumper and patch cord lengths should not exceed 20 meters (66 ft).

Telecommunications equipment that connects directly to main or intermediate cross-connects should do so via cables of 30 meters (98 ft) or less.

Grounding shall meet the appropriate requirements and practices of applicable authorities or codes. Additionally, grounding/bonding shall conform to ANSI/TIA-607-C.

Twisted-pair intra-building backbone cables typically consist of 25 or more pairs. These cables generally have pair counts of 25, 50, 100, 150, 200, 300, and up to a maximum of 600 pairs without a corrugated aluminum outer shield and up to 1,800 pairs with a corrugated aluminum outer shield. These cables may be UL listed communications riser (CMR) or communications plenum (CMP) cable.

The shield of multi-pair UTP backbone cables with a corrugated aluminum shield must be grounded at any point where the outer jacket is removed. Normally, these cables are placed point-to-point between an equipment room and a telecommunications room. At each end, the jacket and shield is removed to expose the twisted pairs for termination.

A bonding clamp must be installed on both ends of the cable and a ground wire placed between the bonding clamp and the bonding busbar using a 6 AWG or larger wire.

Multi-pair UTP backbone cables are constructed by arranging the appropriate number of 25-pair binder groups into a core. The binder groups are then covered with a metallic shield if needed and an outer jacket. Each of the 25-pair binder groups are identical and use the same color code. **See Figure 3-25.** Each of the 25-pair binder groups is wrapped with a pair of plastic binder ribbons, which are color-coded to identify the binder group number. The color code for binder group numbers follows the same pattern as the color code for pair numbers. **See Figure 3-26.**

UTP backbone cables are terminated on terminal blocks in a fashion similar to 4-pair horizontal cables. The difference is that each row of a 110-type block or each column of a 66-type block is used for terminating a 25-pair unit, rather than six 4-pair cables.

High Pair-Count Copper Telephone Cable

The introduction of plastic insulated conductors (PIC) ushered in the use of today's tip and ring color code. Before plastic, phone companies used the old red (ring) and green (tip) color code for each pair and "rang" out pairs either before or after splicing the cable together. It was very time-consuming to splice high pair-count cable to ring out 1,200 to 1,800 pairs, for example. Color coding the pairs meant quick identification and did not require the extra time to identify the correct pair by ringing them out.

In the beginning, several color combinations were attempted, but using 10 very different colors was soon agreed upon. Five would identify the tip colors and five would identify the ring colors. As a result, the tip colors are white, red, black, yellow, and violet; the ring colors are blue, orange, green, brown, and slate. This allows for 25 color combinations that can easily be multiplied for high pair-count cables. **See Figure 3-27.**

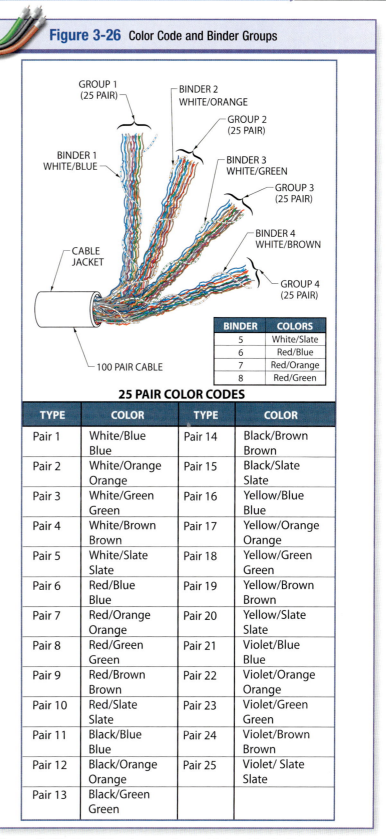

Figure 3-26 Color Code and Binder Groups

BINDER	COLORS
5	White/Slate
6	Red/Blue
7	Red/Orange
8	Red/Green

25 PAIR COLOR CODES

TYPE	COLOR	TYPE	COLOR
Pair 1	White/Blue Blue	Pair 14	Black/Brown Brown
Pair 2	White/Orange Orange	Pair 15	Black/Slate Slate
Pair 3	White/Green Green	Pair 16	Yellow/Blue Blue
Pair 4	White/Brown Brown	Pair 17	Yellow/Orange Orange
Pair 5	White/Slate Slate	Pair 18	Yellow/Green Green
Pair 6	Red/Blue Blue	Pair 19	Yellow/Brown Brown
Pair 7	Red/Orange Orange	Pair 20	Yellow/Slate Slate
Pair 8	Red/Green Green	Pair 21	Violet/Blue Blue
Pair 9	Red/Brown Brown	Pair 22	Violet/Orange Orange
Pair 10	Red/Slate Slate	Pair 23	Violet/Green Green
Pair 11	Black/Blue Blue	Pair 24	Violet/Brown Brown
Pair 12	Black/Orange Orange	Pair 25	Violet/ Slate Slate
Pair 13	Black/Green Green		

Figure 3-26. Each 25-pair group in a cable has identically-colored wire pairs, but the binder is uniquely colored to differentiate between groups. Note: The first color is the insulation color, and the second color is the marking color.

Figure 3-27 Tip and Ring Color Code

TIPS		RINGS	
White	W	Blue	Bl
Red	R	Orange	O
Black	Bk	Green	G
Yellow	Y	Brown	Br
Violet	V	Slate	S

Figure 3-27. The first five pairs use white as the tip color.

Figure 3-28 Pairs 6 through 10

Pair	Tip-Ring	Abbreviation
6	Red-Blue	R-Bl
7	Red-Orange	R-O
8	Red-Green	R-G
9	Red-Brown	R-Br
10	Red-Slate	R-S

Figure 3-28. The second five pairs use red as the tip color.

Tip and Ring Colors

When referring to a wire pair, proper practice is to refer to the tip color first and the ring color second; for example, white-blue (white tip, blue ring) or red-orange (red tip, orange ring). The reason for this is to create consistency in splicing cables. This in turn helps to eliminate identification errors.

The first five pairs (1-5) use the white tip color twisted together with the five ring colors used in order. The second five pairs (6-10) use the next tip color, red, and the five ring colors again in order. **See Figure 3-28.**

This pattern continues, using each tip color with each of the ring colors until 25 cable pairs have been identified. When placed in a multi-pair cable, this group of 25 pairs is known as a *binder unit* or *binder group*.

Binder Groups

For cables larger than 25 pairs, each binder group of 25 pairs is identified by binding the pairs in colored ribbon. **See Figure 3-29.**

Figure 3-29 Binder Groups

Binder 1	White/Blue ribbon	Pairs 1-25
Binder 2	White/Orange ribbon	Pairs 26-50
Binder 3	White/Green ribbon	Pairs 51-75
Binder 4	White/Brown ribbon	Pairs 76-100
Binder 5	White/Slate ribbon	Pairs 101-125
Binder 6	Red/Blue ribbon	Pairs 126-150

Figure 3-29. Binder group ribbons identify which pairs are associated.

Typically, pair colors are separated with a – (dash) and binder group colors are separated with a / (slash); Pair 5 is W-S and Binder 5 is W/S.

This binder group identification system works for up to a 600-pair cable, with binder group 24 containing pairs 576-600 and a binder color of V/Br. Binder numbers are from one through 24, so there will never be a Violet/Slate binder color. **See Figures 3-30 and 3-31.**

To continue on to higher pair count cables, *Super units* are used. Super units start with the 25th binder group (Pairs 601-625). Super units are created by taking the first 24 binders and wrapping them with an additional white marker. An additional red marker will identify the next super unit, which starts with the

Tech Fact

25-Pair Color Code Matrix

From this chart, the following observations can be made:

1) Ones and sixes always have blue rings.
2) Twos and sevens always have orange rings.
3) Threes and eights always have green rings.
4) Fours and nines always have brown rings.
5) Zeros and fives always have slate rings.

TIPS	RINGS				
	Blue	Orange	Green	Brown	Slate
White	1	2	3	4	5
Red	6	7	8	9	10
Black	11	12	13	14	15
Yellow	16	17	18	19	20
Violet	21	22	23	24	25

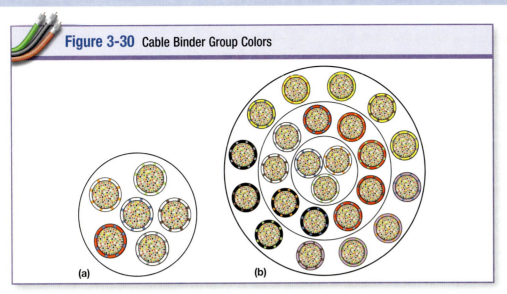

Figure 3-30 Cable Binder Group Colors

(a)   (b)

Figure 3-30. *Binders are used to differentiate between 25-pair groups in cables larger than 25 pairs, such as 150-pair (a) or 600-pair (b) cables. The binder groups are twisted together and covered in an overall sheath.*

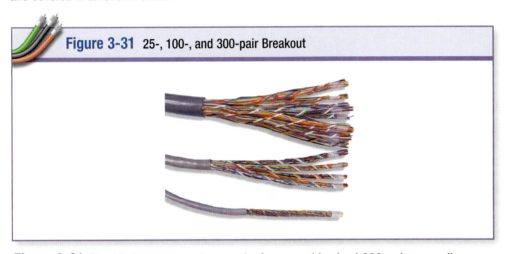

Figure 3-31 25-, 100-, and 300-pair Breakout

Figure 3-31. *The binder group system works for any cable sized 600-pair or smaller.*

25th binder group (W/Bl) and includes each additional binder group until another 600-pairs are reached (pairs 601 to 1,200 are in the red super unit). An additional black marker will surround the 49th binder group (pairs 1,201-1,225) and will continue up to pair 1,800. **See Figure 3-32.**

CORDS AND ADAPTERS

The work area components extend from the telecommunications outlet (connector end of the horizontal cabling system) to the station equipment. The station equipment can be any of a number of devices, including, but not limited to telephones, data terminals, and computers. Work area

cabling is critical to a well-managed distribution system. However, it is generally not permanent and is designed so that it is relatively easy to change. Work area cabling typically consists of cords and adapters. **See Figure 3-33.**

Each 4-pair cable shall be terminated in an 8-position modular jack at the work area. The maximum horizontal cable length has been specified with the assumption that a maximum length of five meters (16 ft) of patch cord has been used in the work area. Cables and connectors used in the work area shall meet or exceed patch cord performance requirements in ANSI/TIA-568.2-D and ANSI/TIA-568.3-D.

Figure 3-32 Super Units

Super Unit Color	Pairs	Binders
White Super Unit	1-100	W/Bl, W/O, W/G, W/Br
	101-200	W/S,R/Bl, R/O, R/G
	201-300	R/Br, R/S, Bk/Bl, Bk/O
	301-400	Bk/G, Bk/Br, Bk/S, Y/Bl
	401-500	Y/O, Y/G, Y/Br, Y/S
	501-600	V/Bl, V/O, V/G, V/Br
Red Super Unit	601-700	W/Bl, W/O, W/G, W/Br
	701-800	W/S,R/Bl, R/O, R/G
	801-900	R/Br, R/S, Bk/Bl, Bk/O
	901-1000	Bk/G, Bk/Br, Bk/S, Y/Bl
	1001-1100	Y/O, Y/G, Y/Br, Y/S
	1100-1200	V/Bl, V/O, V/G, V/Br
Black Super Unit	1201-1300	W/Bl, W/O, W/G, W/Br
	1301-1400	W/S,R/Bl, R/O, R/G
	1401-1500	R/Br, R/S, Bk/Bl, Bk/O
	1501-1600	Bk/G, Bk/Br, Bk/S, Y/Bl
	1601-1700	Y/O, Y/G, Y/Br, Y/S
	1701-1800	V/Bl, V/O, V/G, V/Br

Figure 3-32. Super units are used to organize larger cable assemblies.

Work area cabling may vary in form depending on the application. A cord with identical connectors on both ends is commonly used. When application-specific adaptations are needed at the work area, they shall be external to the telecommunications outlet/connector.

The following installation requirements shall be followed to ensure acceptable link and channel performance.

- Cabling shall comply with applicable codes and regulations.
- Cable and components shall be visually inspected for proper installation.
- Cable stress such as that caused by tension in suspended cable runs and tightly cinched bundles should be minimized.
- Cable ties used to bundle cables should be applied loosely to allow the cable tie to slide around the cable bundle, and should not be cinched so tightly as to deform the cable sheath.
- Cable placement should not deform the cable sheath.

Figure 3-33 Cords and Adapters

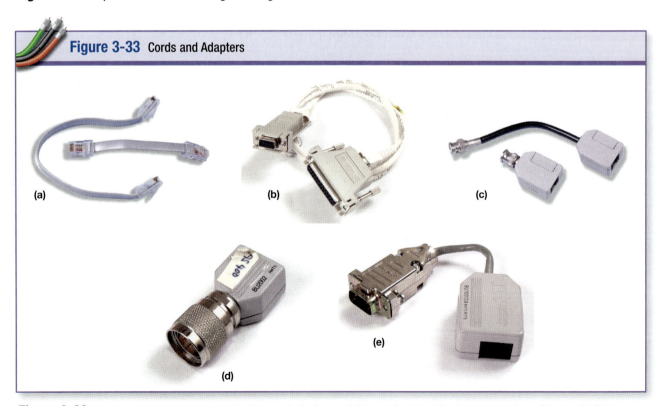

(a)

(b)

(c)

(d)

(e)

Figure 3-33. Commonly used cords and adapters include modular cords (a) and adapters such as Pin to 8P8C (b), Coax to 8P8C (c), Twinaxial to 8P8C (d), and Token Ring to UTP (e).

- The minimum bend radius, under no-load conditions, for 4-pair UTP cable shall be four times the cable diameter.
- The maximum pulling tension of 4-pair 24 AWG UTP cable shall be 110 N (25 lbf).
- Cables should be terminated with connecting hardware of the same category or higher.
- Patch cords should be of the same performance category or higher as the horizontal cables to which they connect.
- To maintain the cable geometry, remove the cable sheath only as much as necessary to terminate the cable pairs on the connecting hardware. The connecting hardware manufacturer's instructions for cable sheath strip-back shall be followed.
- When terminating category 5 and higher cables, the cable pair twists shall be maintained to within 13 millimeters (0.5 in) from the point of termination. When terminating category 3 cables, the cable pair twists shall be maintained to within 76 millimeters (3 in) from the point of termination. For best performance, when terminating cable on connecting hardware, the cable pair twists should be maintained as close as possible to the point of termination.
- The minimum bend radius for multi-pair cable shall be ten times the cable diameter.
- For multi-pair cable, manufacturer's pulling tension guidelines shall be followed.
- The applied bend radius for intra-building 2- and 4-fiber horizontal optical fiber cable shall not be less than one inch under no-load conditions. When under a maximum tensile load of 222 N (50 lb/ft), the applied bend radius shall not be less than two inches.
- The applied bend radius for intra-building optical fiber backbone cable shall not be less than that recommended by the manufacturer. If no recommendation is known, the applied bend radius shall not be less than 10 times the cable outside diameter under no-load conditions and not less than 20 times the cable outside diameter when the cable is under tensile load.
- The applied bend radius for inter-building optical fiber backbone cable shall not be less than that recommended by the manufacturer. If no recommendation is provided or known, the applied bend radius shall not be less than 10 times the cable outside diameter under no-load conditions and not less than 20 times the cable outside diameter when the cable is under a tensile load up to the rating of the cable, usually 600 lb/ft.
- Pathways and spaces in support of horizontal and backbone cabling shall be designed and installed in accordance with the TIA-569-E Standard.
- Telecommunications rooms, equipment rooms, and entrance facilities shall be designed and installed in accordance with the TIA-569-E Standard.
- A bonding and grounding system conforming to the requirements of the TIA-607-C Standard shall be installed.

AN INTRODUCTION TO FIBER OPTIC CABLING SYSTEMS

Although most structured cabling systems consist of copper cabling, many systems also utilize fiber optic cabling; this is especially true in the cabling backbone. Similar to copper conductors and cables, fiber optic conductors and cables are used to transmit data from one location to another location. Fiber optics is the method of using light to transport information from one location to another location through thin filaments of glass or plastic. In a fiber optic system, electrical information (video, voice, or data) is converted to light energy by a fiber optic transceiver, transmitted to a receiver over optical fibers, and then converted back into electrical information by another fiber optic transceiver. Fiber optic cabling system performance is defined by param-

The Other *IDC Terminal Blocks*

There are four basic types of insulation displacement connector (IDC) blocks. The two most often specified for commercial building use are the 66-type block and the 110-type block. However, there are two more that, while not now commonly specified, are still available and still in widespread use. They are the BIX™ and the LSA (also known as KRONE or KRONE Blocks).

BIXTM (Building Industry Cross-Connect)

BIX-type hardware was originally a product of Nortel Networks™ and was developed in the 1970s. The BIX termination blocks support 25-pair cabling and look similar to 110-type blocks. However, where 110-type blocks place "C" type connecting blocks on top of the wiring block, BIX-blocks are of a pass-through design. The BIX block has greater distance between adjacent contacts for better crosstalk performance. BIX mount assemblies come in 50-pair, 250-pair, 300-pair, and 900-pair.

BIX-type blocks are terminated, then flipped over in their mount to expose the other end for cross-connect duty.

BIX hardware is intended for use in terminating and cross-connecting telephone system conductors.

A 25-pair cable is tie-wrapped to the BIX block while in its mount on the right side of the block. Pairs are laid out and terminated as on a 110-type block, starting at position 1 on the left side of the block. After dressing the pairs and punching them down, the block is flipped over to expose the other side.

The BIX tool uses a scissor action to cut the wires. BIX-type hardware supports category 5e installations. BIX-type hardware may now be found with a Belden representative.

BIX™ style IDC terminal blocks and mount

Individual BIX™ block

Slot for Inserting a Tie Wrap

Tie wrap slots on BIX™ blocks

BIX™ tool

eters similar to those used to define copper cabling. The two main types of fiber optic cables include multi-mode and single-mode. **See Figure 3-34.**

Fiber optic cables are installed in both premise networks and public networks. A *premise network* is a transmission network inside a building or group of buildings that connects various types of voice and data communications devices, switching equipment, and other information management systems to each other and to outside communications networks. Examples of premise networks include office buildings, school campuses, hospitals, and casinos. A *public network* is a network operated by common carriers for the administration

The Other IDC Terminal Blocks (cont.)

The KRONE block (now a division of CommScope) was developed in Germany in the late 1970s, and features silver-plated contacts positioned at a 45° angle to the wire. The KRONE style block is still very popular with local exchange carriers (LECs) for their outside plant distribution and delivery of broadband services, such as DSL. One application for the KRONE type termination hardware is in fiber to the node (FTTN) broadband delivery. This is where a fiber optical carrier is brought to a cabinet containing a digital subscriber line access multiplexer (DSLAM) and where DSL signals are sent out over existing copper to the subscriber.

KRONE style termination blocks

KRONE blocks have silver-plated contacts placed at 45° to the wire.

1-pair test cord and 4-pair to RJ45 patch cord

The KRONE block offers a port between the incoming and outgoing cable that enables monitoring or redirecting of individual circuits. The ports also allow for the temporary or permanent disconnect of a circuit.

KRONE test cord and 4-pair patch cord pin-out

of circuits for use by the public. Examples of public networks are telephone, Internet, and cable television systems.

Optical fiber can be used for telecommunications and networking because it is flexible and it can be bundled as cables. Fiber optic technology is advantageous for long-distance communications because light travels through the fiber with little attenuation compared to copper cables. Additionally, each fiber can carry many independent channels with a different wavelength of light (wavelength-division multiplexing). Over short distances, such as networking within a building, fiber saves space in cable ducts because a single fiber can carry much more data than a single copper cable. Fiber is also immune

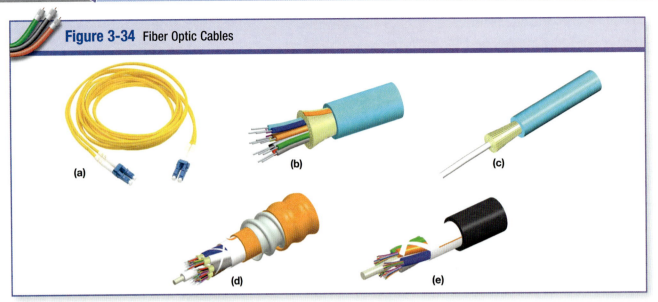

Figure 3-34 Fiber Optic Cables

Figure 3-34. *Optical fibers come in many cabling configurations, including: (a) single mode (OS2) duplex patch cable, (b) 12 strand multi-mode (50 micron OM 3) distribution cable, (c) multi-mode (50 micron) simplex cable, (d) multi-mode (62.5 micron, OM2) armored distribution cable, and (e) outdoor loose tube cable.*

to electrical interference, which prevents crosstalk between signals in different cables and pickup of environmental noise. Fiber cabling is more secure than copper cabling because wiretapping into fiber is more difficult than copper. Because they are non-electrical, fiber cables are unaffected by high electrical potential differences and can be used in environments where explosive fumes are present without danger of ignition.

Although fibers can be made out of transparent plastic or glass, the fibers used in long-distance telecommunications applications are always glass because of the lower optical attenuation. Both multi-mode and single-mode fibers are used in communications, with multi-mode fiber used mostly for distances up to 500 meters, and single-mode fiber used for longer distance links.

To protect a fiber strand, a protective coating surrounds the glass; to protect the cable, additional strength members are used. The transparent glass cladding surrounds the core with a lower index of refraction, providing total internal reflection of the propagating signal. This contains the light signal within the core and protects against light scattering. **See Figure 3-35.** All fibers have a cladding diameter of 125 micrometers. **See Figure 3-36.** The core diameters vary according to the type of fiber. For example, 62.5-micrometer multi-mode fiber cabling is the old standard for premises networks in North America, 8- to 10-micrometer single-mode fiber is used for public networks and large premise networks, and 50-micrometer multi-mode laser optimized fiber is the new standard for premises (commercial offices and institutional campuses) used to support 10 gigabit per second (Gb/s) data systems.

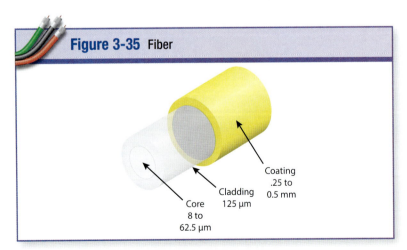

Figure 3-35 Fiber

Core
8 to
62.5 μm

Cladding
125 μm

Coating
.25 to
0.5 mm

Figure 3-35. *Fiber, a thin filament of glass or plastic that conducts a light signal, is made of dielectric material consisting of core and cladding, which allows total internal reflection of light for propagation.*

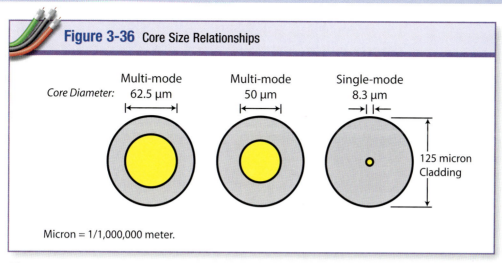

Figure 3-36 Core Size Relationships

Core Diameter:
Multi-mode 62.5 µm
Multi-mode 50 µm
Single-mode 8.3 µm

125 micron Cladding

Micron = 1/1,000,000 meter.

Figure 3-36. *In order to increase transmission rates, single-mode fiber has a much smaller core diameter than multi-mode.*

Multi-Mode Fiber

A *multi-mode fiber (MM)* is an optical fiber that can carry multiple signals (distinguished by frequency or phase) at the same time. It is a graded or step-index optical fiber that supports the propagation of more than one light mode. Although multi-mode fiber cabling has typically been installed in premises networks, more single-mode fiber is being installed for the same purpose. A *refractive index* is the ratio of the velocity of light in the core and cladding of the same fiber. Multi-mode fibers can have graded index cores, where the refractive index has a gradual rather than abrupt change from the inner core to the outer core (stepped index), at which point the refractive index matches that of the cladding. Because the light path is more elliptical in shape, graded index cores are an improvement over stepped index cores. **See Figure 3-37.**

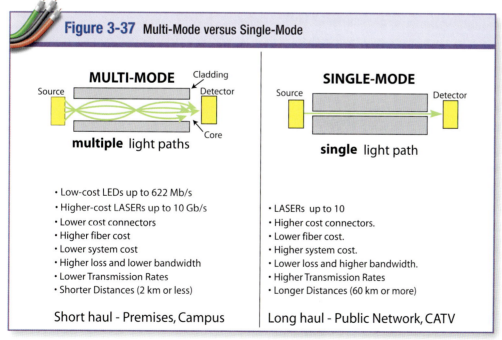

Figure 3-37 Multi-Mode versus Single-Mode

MULTI-MODE

Source
Cladding
Detector
Core

multiple light paths

SINGLE-MODE

Source
Detector

single light path

- Low-cost LEDs up to 622 Mb/s
- Higher-cost LASERs up to 10 Gb/s
- Lower cost connectors
- Higher fiber cost
- Lower system cost
- Higher loss and lower bandwidth
- Lower Transmission Rates
- Shorter Distances (2 km or less)

Short haul - Premises, Campus

- LASERs up to 10
- Higher cost connectors.
- Lower fiber cost.
- Higher system cost.
- Lower loss and higher bandwidth.
- Higher Transmission Rates
- Longer Distances (60 km or more)

Long haul - Public Network, CATV

Figure 3-37. *The two main types of fiber available for fiber optic cabling are multi-mode and single-mode fiber.*

Advantages of multi-mode fiber include low-cost LEDs, connectors, and systems. Disadvantages of multi-mode fiber include high-cost fiber, high insertion loss, low transmission rates, and installation distances of up to only two kilometers.

Single-Mode Fiber

A *single-mode fiber (SM)* is an optical fiber where a signal travels along one path. Single-mode fiber cabling is prevalent in service provider networks because its lower loss and higher bandwidth allow transmission of optical signals at extremely high data rates over long distances using laser sources. Single-mode fiber cabling (8- to 10-micrometer core diameter) is primarily used in service-provider (telephone, Internet, and cable television) networks. However, some premises networks utilize single-mode fiber cabling when distances of two kilometers or greater are required.

Advantages of single-mode fiber include lower fiber cost, low insertion loss, high bandwidth capabilities, high transmission rates, and use with distances of 60 kilometers and greater. Disadvantages of single-mode fiber include high-cost laser transmitters, connectors, and high system cost.

Fiber Optic Characteristics

The two main considerations that affect data transmission for fiber optic cabling systems are attenuation and bandwidth. Attenuation, as related to optical fiber, is optical power loss. Bandwidth is a measure of the information-carrying capacity of a cabling system. In a fiber optic system, as wavelength increases (or frequency decreases), attenuation decreases.

Attenuation is the reduction of optical power as it propagates through an optical fiber. Attenuation occurs in the fiber itself because of two phenomena: absorption and scattering. Scattering is the most common form of attenuation in optical fibers. As the name implies, scattering means that the light mode(s) traveling through a fiber are scattered in all directions. This is caused by impurities in the fiber and by collisions with the actual glass molecules. This is known as *Rayleigh Scattering*.

Absorption is the second most significant cause of attenuation in an optical fiber. Absorption is caused by the collision of photons with impurities or dopants in the glass. These impurities are primarily OH- ions (negatively charged water molecules). The dopants are added to the glass to make it more flexible. **See Figure 3-38**.

Fortunately, these two sources of attenuation are linear and predictable. A spool of cable will have the same amount of attenuation along its entire length. Optical fiber is specified with an attenuation coefficient that indicates the total attenuation per kilometer of length. **See Figure 3-39**.

---

50 μm Laser-Optimized Fiber

In 1996, computer networking and cabling groups began discussing the need for multi-mode fiber cabling to support data rates higher than one Gb/s. These high data rates require the use of optical transmitters with faster modulation capabilities than those of LEDs. Devices such as the vertical cavity surface emitting laser (VCSEL) or the edge-emitting semiconductor laser diode are used as optical sources in these higher-speed systems. Unlike the less coherent LEDs, laser transmitters typically excite only a subset of the fiber modes available. This requires compatible specifications on both the transmitter launch conditions and the fiber modal dispersion properties to support these high-speed systems. After much testing, industry experts agreed that a high-performance multi-mode fiber with a core diameter of 50 micrometers was the most cost-effective cabling solution for these higher-speed systems.

In 2002, an addendum to the ANSI/TIA/EIA-568-B.3 standard was published. The addendum (568-B.3-1), titled *Additional Transmission Performance Specifications for 50/125 μm Optical Fiber Cables*, specifies requirements for a 50/125 micrometer optical fiber cable capable of supporting 10 Gb/s transmission at a distance of 300 meters (984 ft) using 850-nanometer wavelength lasers. These laser-optimized fibers are allowed to have an attenuation rate of up to 3.5 decibels per kilometer and must have a laser bandwidth of at least 2,000 megahertz per kilometer at a wavelength of 850 nanometers. Laser-optimized fibers typically have a light blue or aqua-colored jacket.

Figure 3-38 Attenuation by Wavelength

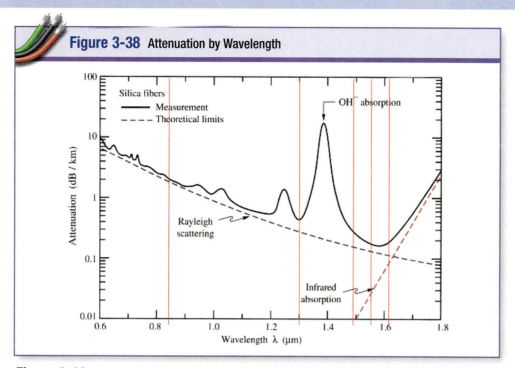

Figure 3-38. *Absorption, a major source of attenuation, varies with wavelength.* Courtesy of Ideal, Inc.

Figure 3-39 Parameters of Multi-Mode and Single-Mode Fiber

Parameter	Multi-Mode	Single-Mode
Operating Wavelengths (λ)	850/1300 nm	1310/1550/1625 nm
Fiber Cladding Diameter	125 µm	125 µm
Fiber Core Diameter	50 or 62.5 µm	8-10 µm
Typical Fiber Bandwidth	500/500 MHz/km for 50 µm 200/500 MHz/km for 62.5 µm	Depends on MFD
Typical Attenuation Coefficient	3.0 dB/km @ 850 nm 1.5 dB/km @ 1300 nm	0.4 to 1.0 dB/km @ 1310 nm 0.4 to 1.0 dB/km @ 1550 nm
Maximum Connector Insertion Loss	0.75 dB/connector pair	0.75 dB/connector pair
Maximum Splice Insertion Loss	0.30 dB each	0.30 dB each
Minimum Return Loss for Connectors or Splices	20 dB	35 dB 55 dB for analog video (CATV)
Light Source	LED/VCSEL	Laser
Type of Dispersion	Modal & Chromatic	Chromatic only
Application	Short Range Backbone/LAN	Long Range Backbone/Telco/CATV

Figure 3-39. *Fiber optic parameters can be found in the ANSI/TIA-568.3-D Standard.* Courtesy of Ideal, Inc.

Attenuation

The ANSI/TIA-568.3-D Standard contains fiber attenuation limits. This Standard allows an attenuation of up to 0.75 decibels per connection for connectors and up to 0.3 decibels for either mechanical or fusion splices. Additional requirements in the ANSI/TIA-568.3-D Standard for measurement of fiber attenuation include:

- Measurements must be made at one wavelength for horizontal cabling
- Measurements must be made at two wavelengths for backbone cabling

Bandwidth

Fiber manufacturers measure the modal bandwidth of multi-mode fibers using laser sources at wavelengths of 850 and 1,300 nanometers and using equipment and procedures that conform to relevant TIA standards. Several kilometers of fiber are wound on plastic spools. The manufacturer measures the bandwidth (in MHz) and divides by the fiber's length (in km). The fiber is reported as a bandwidth per distance product (MHz/km). This megahertz per kilometer (MHz/km) product is a figure that can be used to compare the bandwidth quality of various fibers. For example, the ANSI/TIA-568.3-D Standard requires a minimum bandwidth of 1,500 MHz/km at a wavelength of 850 nanometers, and a minimum bandwidth of 500 MHz/km at a wavelength of 1,300 nanometers for multi-mode fibers. Because bandwidth cannot be adversely affected by installation procedures, there are no field test requirements for it.

FIBER OPTIC BACKBONE CABLE

A typical intra-building fiber optic backbone cable contains 12 to 72 fibers. Each fiber is covered with a color-coded buffer 900 micrometers in diameter. **See Figure 3-40.**

When cables contain more than 12 fibers, the fibers are generally separated into units of six or 12 fibers, and placed in colored PVC tubes within the outer jacket of the cable.

FIBER OPTIC TERMINATION

Fiber optic cables are terminated in cabinets or shelves located in TRs and ERs. After a few feet of cable jacket are removed, the cable is secured, terminated,

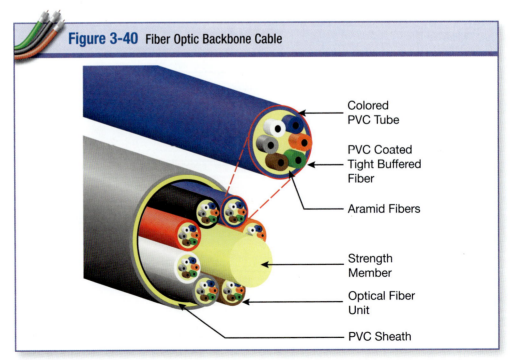

Figure 3-40 Fiber Optic Backbone Cable

Colored PVC Tube

PVC Coated Tight Buffered Fiber

Aramid Fibers

Strength Member

Optical Fiber Unit

PVC Sheath

Figure 3-40. Fiber optic backbone cables can contain between 12 and 72 fibers.

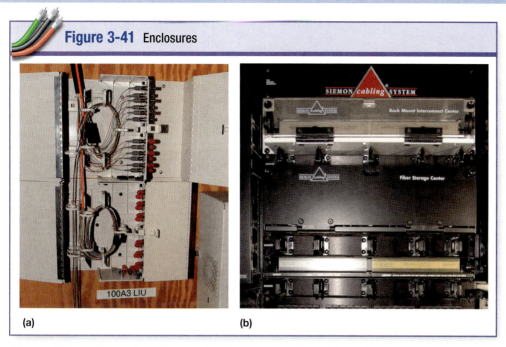

Figure 3-41 Enclosures

100A3 LIU

(a)   (b)

Figure 3-41. Fiber optic cables are terminated in cabinets (a) or shelves (b) located in equipment rooms and/or telecommunications rooms.

and the connectors are plugged into a coupler panel. **See Figure 3-41.**

Fiber Optic Connectors
A fiber optic connector must be terminated on each end of a fiber. The connector may be an epoxy style, pigtail, or pre-terminated (crimp on) type. Epoxy style connectors require field polishing of the connector end face; a pigtail requires fusion splicing. Pre-terminated connectors do not require field polishing; however, they will typically exhibit higher loss than the other styles of connectors. **See Figure 3-42.**

Installation Requirements
To ensure a connection will maintain the correct polarization throughout the cabling system, the correct adapter orientation and optical fiber color code shall be followed. **See Figure 3-43.** Once the system is installed and correct polarization is verified, the optical fiber cabling system shall maintain the correct polarization of transmit and receive fibers and will not be a concern for the end user.

The ability to support polarization is one of the major advantages of the SC and LC connectors. For the other types of connectors, polarization may cause

serious problems with administration. The best way to verify correct polarization for the transmitter to receive is to use an optical continuity checker and a power meter. A continuity checker allows one to verify cable ends by observing light. The power meter will indicate the transmitter cable by displaying a decibel per meter (dB/m) reading.

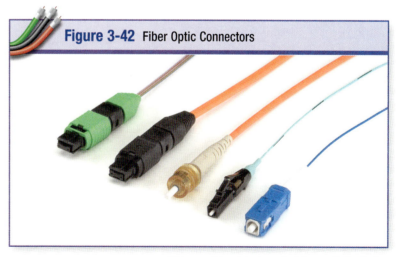

Figure 3-42 Fiber Optic Connectors

ST is a registered trademark of Lucent Technologies.

Figure 3-42. From left to right: MPO connector ribbon; MPO connector, distribution; ST style connector; LC style connector; SC style connector.
Courtesy of Corning, Inc.

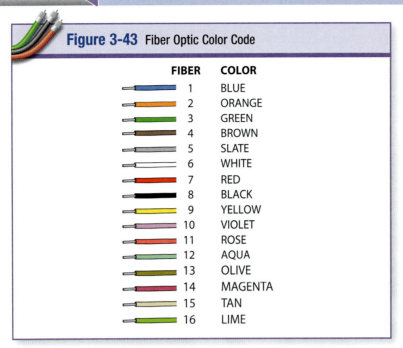

Figure 3-43 Fiber Optic Color Code

FIBER	COLOR
1	BLUE
2	ORANGE
3	GREEN
4	BROWN
5	SLATE
6	WHITE
7	RED
8	BLACK
9	YELLOW
10	VIOLET
11	ROSE
12	AQUA
13	OLIVE
14	MAGENTA
15	TAN
16	LIME

Figure 3-43. Cables with 12 or multiples of 12 fibers will use the first 12 colors, adding tracers for subsequent groups of 12. Array connectors using 16 fibers use the color code as shown.

Fiber Optic Polarity

Each fiber optic cabling segment shall be installed in such a way that odd-numbered fibers are Position A at one end and Position B at the other end, while the even-numbered fibers are Position B at one end and Position A at the other end.

Fiber optic patch cords shall be 2-fiber patch cords of the same fiber type as the optical channel. Fiber optic patch cords shall be configured so that A connects to B and B connects to A (transmit connects to receive and receive connects to transmit). **See Figure 3-44.**

Optical Fiber Cabling

The 568SC implementation shall be achieved by using consecutive fiber numbering (for example, 1,2,3,4...) on both ends of an optical fiber link, but the 568SC adapters shall be installed in opposite manners on each end (for example, A-B, A-B... on one end, and B-A, B-A... on the other). **See Figure 3-45.**

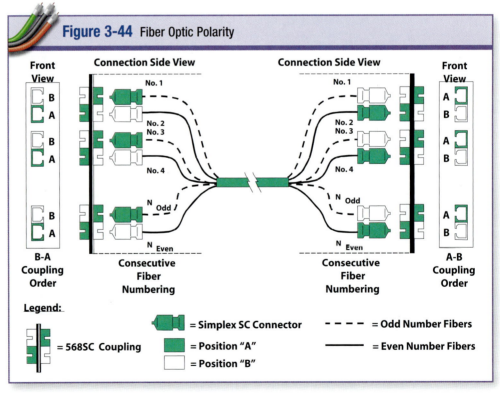

Figure 3-44 Fiber Optic Polarity

Figure 3-44. Odd- and even-numbered fibers must be installed at different ends of the connection.

For other duplex connector styles, the objective may be achieved either by the above method, or by using reverse-pair positioning. Reverse-pair positioning is achieved by installing fibers in consecutive fiber numbering (for example, 1,2,3,4 ….) on one end of an optical fiber link and reverse-pair numbering (for example, 2,1,4,3….) on the other end of the optical fiber link.

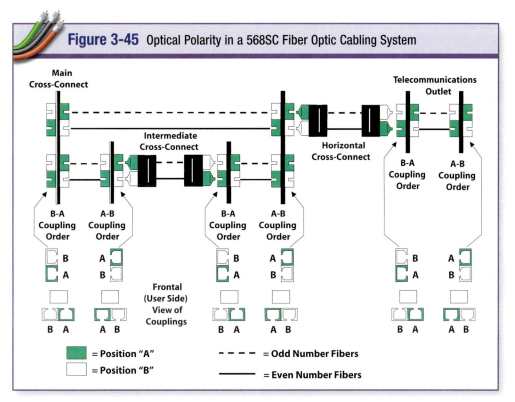

Figure 3-45 Optical Polarity in a 568SC Fiber Optic Cabling System

Figure 3-45. The 568SC configuration is achieved by using consecutive fiber numbering and installing the adapters in opposite manners on each end.

SUMMARY

Although it may seem complicated at first, a structured cabling system really narrows the choices as to what cables and connectors to use. Most horizontal cabling, for example, uses UTP. A complete end-to-end horizontal UTP cabling system requires an 8P8C modular outlet, less than 90 meters of UTP cable, a modular patch panel of some type and/or a rated cross connect (terminal block), and a couple of patch cords, bringing the entire horizontal cable length to no more than 100 meters.

Installers must become familiar with each of the horizontal cabling types and their connecting hardware. These are 4-pair 100-ohm unshielded or shielded twisted-pair cabling: category 5e, category 6, or category 6A; multi-mode optical fiber cabling; and single-mode optical fiber cabling.

Backbone cabling requires familiarity with high pair-count cables, multi-mode optical fiber cabling, and single-mode optical fiber cabling. Most new installations of backbone cable will likely be single-mode optical fiber because of its ability to carry large amounts of data, its noise immunity, and its security. However, the installer will still come across many installations that use high pair-count copper and/or multi-mode optical fiber. In each case, the installer will need to be familiar with the installation techniques of the existing cable plant.

REVIEW QUESTIONS

1. The maximum horizontal cabling distance shall be __?__, independent of media type.
   a. 27.4 m (90 ft)
   b. 90 m (295 ft)
   c. 99 m (325 ft)
   d. 134 m (440 ft)

2. The ANSI/TIA __?__ standard requires horizontal cabling to be implemented in a star topology.
   a. 568.0-D
   b. 568.1-D
   c. 568-C.2
   d. 568-3.D

3. The 8P8C modular outlet is only compatible with 4-pair circuits.
   a. True
   b. False

4. For category 5e and higher cables, pair twist must be maintained to within __?__ of the termination.
   a. 6.5 mm (0.25 in)
   b. 13 mm (0.5 in)
   c. 19 mm (0.75 in)
   d. 76 mm (3 in)

5. Which *NEC* article covers type OFC cable?
   a. 725
   b. 760
   c. 770
   d. 800

6. How much outside plant cable entering a building is allowed to be run exposed inside the building?
   a. 7.5 m (25 ft)
   b. 15 m (50 ft)
   c. 22.5 m (75 ft)
   d. 30 m (100 ft)

7. In a 25-pair cable, which pair is represented by the colors Red/Orange – Orange?
   a. Pair 4
   b. Pair 5
   c. Pair 6
   d. Pair 7

8. Binder 5 (White/Slate ribbon) represents pairs __?__.
   a. 26-50
   b. 51-75
   c. 101-125
   d. 126-150

9. Which super unit color represents pairs 1,201-1,800?
   a. Black
   b. Red
   c. Violet
   d. White

10. The maximum bend radius, under no-load conditions, for 4-pair UTP cable shall be eight times the cable diameter.
    a. True
    b. False

Structured Cabling System Performance

Introduction

Unshielded twisted-pair (UTP) cabling systems, originally used to transmit voice signals, have been around for years. In that time, the performance of UTP has been dramatically improved. Cable manufacturers continue to find ways to squeeze even more performance out of copper UTP and existing modular 8-position/8-conductor (8P8C) outlets and plugs. However, increased performance of UTP cable and hardware comes with its own set of issues.

The installation standards require that cable and connecting hardware meet certain performance specifications. To understand UTP cable performance, one must first look at a single twisted pair's electrical characteristics. Everything builds from there. Understanding the characteristics of UTP copper cable improves troubleshooting skills when testing structured cabling systems.

Objectives

- Explain why twisting a pair reduces its susceptibility to noise and recognize the benefit of using a different twist length for each pair in a 4-pair UTP cable.
- Explain the various structured cabling channel parameters.
- Describe the differences among the standard categories of UTP cables.
- Describe basic field testing parameters.
- Identify types of shielded cables.

Chapter 4

Table of Contents

DATA TRANSFER RATE

Data transfer rate is the speed at which data is transferred from one device to another. This is the rate in bits per second at which information is transferred between network devices over a communications channel. It is sometimes referred to as *throughput* or *operating speed*.

Although data transfer rates can be expressed in megabits or megabytes, electronics standards and products typically specify data rates in terms of megabits per second (Mb/s). *Frequency* is a measure of how often a waveform is completed in relation to time, expressed in cycles per second. The unit of measure is the hertz (Hz). Cabling standards typically specify performance across a range of frequencies in megahertz (MHz). **See Figure 4-1.**

A *bit* is the smallest unit of information that a computer can process. It is a binary digit, the smallest element of information in binary systems. A bit is either a logical one (1) or zero (0), also known as an *on* or an *off* bit, respectively, of binary data. Data is stored in personal computer (PC) memory and in buffers on network interface cards as a series of bits, which are simply logical ones and zeros. Bits are transmitted over one or more twisted pairs in a cable as an encoded digital waveform clocked at a specific rate in megabits per second.

An *encoder* is a device that converts data into code. The particular type of encoder used in a network interface card determines the nature of the transmitted digital signal and determines the required bandwidth for the cabling system. *Bandwidth* is the width of a communications channel. In analog systems, bandwidth is measured in megahertz (cycles per second), while in digital systems, bandwidth is measured in megabits per second. Bandwidth consists of a range of frequencies, usually the difference between the upper and lower limits of the range, expressed in hertz. It is used to denote the potential capacity of the medium, device, or system. In copper and optical fiber cabling, the bandwidth decreases with increasing length.

Given a specific encoder type, as bit rate increases, more bandwidth is required. However, for a given bit rate, when a more complex encoding method is implemented, the bandwidth requirement will be reduced. The bandwidth required for transmission of a digital waveform is determined not only by bit rate, but by the type of encoder that is used with the transmitter. The correlation of megabits to megahertz, therefore, is determined by the specific method of encoding used in a transmitter. In practice, a specific encoding method is selected by a standards committee in order to assure that a newly developed network protocol, or an increased speed for an existing protocol, will operate properly over a particular type of cabling.

Hertz and bit rate are significantly different regarding how they convey information; however, when discussing cable performance, there are terms that can be associated with both. *Signal-to-noise ratio* (*SNR*) is a ratio that expresses the difference between the power level of the information signal being transmitted in a circuit, and the power level of undesired interfering signals that are also present in the circuit (crosstalk, hum, noise, static).

Figure 4-1 Megabits versus Megahertz

Digital Time Domain

1 0 1 0 1 0 1 0   Logical Data

Electrical Representation (Mbps)

Analog Frequency Domain

Amplitude (Volts)

Frequency (MHz)

Spectral Density

Frequencies (MHz)

Figure 4-1. Megabits per second (Mb/s) is a measure for data rates, while megahertz (MHz) measures frequency.

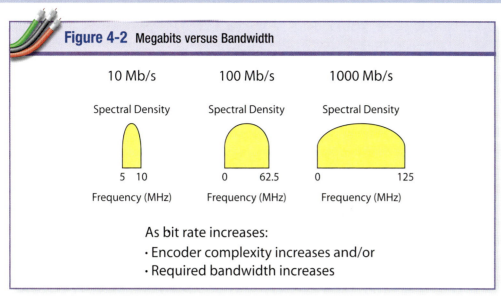

Figure 4-2 Megabits versus Bandwidth

10 Mb/s   100 Mb/s   1000 Mb/s

Spectral Density   Spectral Density   Spectral Density

5 10   0   62.5   0   125

Frequency (MHz)   Frequency (MHz)   Frequency (MHz)

As bit rate increases:
- Encoder complexity increases and/or
- Required bandwidth increases

Figure 4-2. *As the bit rate increases, the need for bandwidth also increases unless a more complex encoding method is used.*

SNR is expressed as a ratio; the larger the number (ratio), the better the circuit quality (information signal power level: undesired interfering signals in the circuit). *Bit error rate (BER)* is the ratio of incorrectly transmitted bits to total transmitted bits. BER, usually expressed as a power of 10, is the number of errors made in a digital transmission as compared to complete accuracy. When the bit power level of a signal gets down close to the noise level of the signal, the decoder could encounter trouble determining if the signal represents a one or a zero. This could cause the decoder to interpret a one as a zero or a zero as a one, causing a bit error.

As bit rate increases, the required bandwidth increases. Bandwidth density can be gauged by the shape of a sine wave as seen on an oscilloscope. The higher the frequency, the larger the bit rate. **See Figure 4-2.**

UNSHIELDED TWISTED-PAIR CABLES

Unshielded twisted pairs are solid copper conductors that are twisted together in a manner that minimizes crosstalk (interference) with other pairs of wires in the same unshielded cable. A UTP cable consists of four pairs; each pair has a ring conductor and a tip conductor. The ring conductor is typically a solid-colored conductor with white stripes, while the tip conductor is white-colored with colored stripes. **See Figure 4-3.**

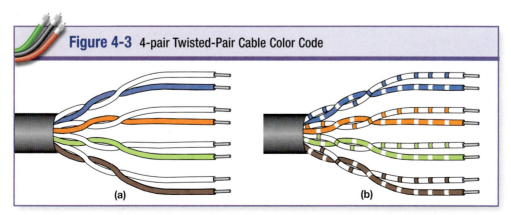

Figure 4-3 4-pair Twisted-Pair Cable Color Code

(a)   (b)

Figure 4-3. *Twisted-pair cabling can use either a solid (a) or striped (b) color code.*

Unshielded twisted-pair cabling systems have been used for decades to carry voice signals. In recent years, tremendous improvements in the performance of UTP cables and outlets have been achieved. Modern equipment can send data signals of up to 10 gigabits per second (Gb/s) over UTP cabling systems at distances of up to 100 meters (328 ft).

As the number of installed UTP cabling systems has increased in recent years, installations of shielded twisted-pair (STP) and coaxial cable systems are decreasing. The reasons for the increase in installations of UTP cabling systems include:

- Smaller cable diameter
- Easier installation
- Support of most voice and data applications
- High performance
- Low cost

Several different parameters affect UTP cable performance, including the physical construction of the cable and the electrical characteristics of the signal transmission. Significant characteristics of cable construction include:

- The conductor material and the conductor diameter
- Insulating material type and thickness of the insulating material
- Spacing between insulated pair conductors as well as spacing between the four UTP pairs
- The number of twists per inch in each of the four pairs of conductors and the twist rate of each of the four pairs in the UTP cable

Electrical characteristics of the signal transmission include:

- Resistance
- Capacitance and capacitive reactance
- Inductance and inductive reactance
- Impedance
- Conductance

Resistance is the physical opposition to electrical current flow. The amount of resistance on a conductor or conductor pair is measured in ohms (Ω) and is determined by the size of the conductor (the larger the conductor, the less resistance it has) and the material of the conductor, most commonly copper. In addition, the insulation type, thickness, and density can factor into the overall resistance of the conductor pair. The overall resistance can cause poor SNR as well as BER.

A *capacitor* is an electric circuit element used to temporarily store a charge. In general, a capacitor consists of two metallic plates separated and insulated from each other by a dielectric. *Capacitance* is the property of a circuit element that permits a charge to be stored. The unit of measure is *farad* and the formula variable is *C*. *Capacitive reactance* is the opposition to alternating current flow caused by a capacitor. Capacitive reactance creates a coupling effect of alternating current signal flow between electrical conductors and components and is represented by the variable X_C. Factors that may result in capacitance and capacitive reactance issues in a cable include crushing, twisting,

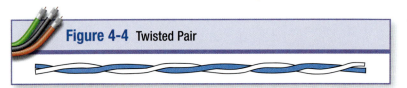

Figure 4-4 Twisted Pair

Figure 4-4. Twisted-pair conductors minimize crosstalk with other pairs in the same cable.

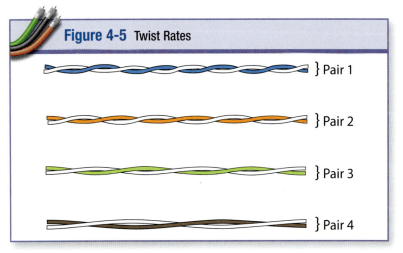

Figure 4-5 Twist Rates

} Pair 1

} Pair 2

} Pair 3

} Pair 4

Figure 4-5. UTP and STP use different twist rates to help reduce noise coupling between pairs.

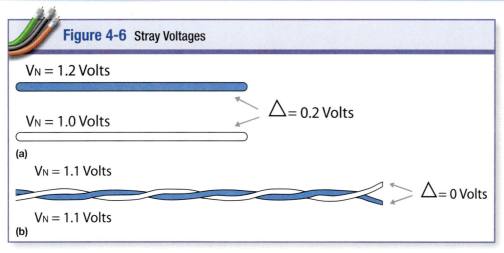

Figure 4-6 Stray Voltages

V_N = 1.2 Volts

V_N = 1.0 Volts

$\triangle$ = 0.2 Volts

(a)

V_N = 1.1 Volts

$\triangle$ = 0 Volts

V_N = 1.1 Volts

(b)

Figure 4-6. Pairs with no twist (a) have varying levels of stray voltage. Twisted pairs (b) eliminate stray voltage.

and bends in the cable that may result in the conductors being closer together or farther apart.

A pair of conductors is twisted in order to improve the electrical balance of a 2-wire circuit so that it will be less susceptible to pick up noise from other conductors in the cable. **See Figure 4-4.** In 25-pair or smaller multi-pair UTP cables, each conductor pair has a different twist length. **See Figure 4-5.** The purpose of different twist lengths is to minimize crosstalk coupling among the pairs. In multi-pair cables larger than 25 pairs, the appropriate number of 25-pair units are bundled together to provide the correct pair count.

Pairs of conductors that are not twisted will be imbalanced with respect to ground. The result of this imbalance is different levels of induced noise voltage on each of the conductors. For example, a receiver connected across the conductors detects a difference in noise voltage between the conductors of 0.2 volts. When a pair is twisted, the electrical characteristics of each conductor with respect to ground are nearly identical, with almost no difference in the induced noise voltage between the two conductors. **See Figure 4-6.**

While twisting a pair of conductors reduces susceptibility to induced voltages from sources external to the cable, a conductor pair's exposure to the coupling of

noise produced by other pairs within the same cable must also be considered. *Capacitive coupling* is the transfer of energy from one circuit to another by virtue of the mutual capacitance between the circuits. If all the pairs within a cable have the same twist length, there can be a great deal of undesirable capacitive coupling among them. When each pair in a UTP cable has a different twist length, the amount of pair-to-pair capacitive coupling is reduced. **See Figure 4-7.**

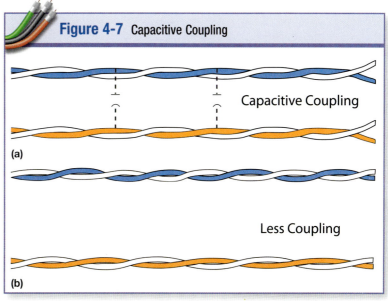

Figure 4-7 Capacitive Coupling

Capacitive Coupling

(a)

Less Coupling

(b)

Figure 4-7. Pairs with the same twist rate (a) may suffer from capacitive coupling. Different twist rates (b) reduce the degree of pair-to-pair coupling.

Inductance is the property of an electric circuit displayed when a varying current induces an electromotive force in a circuit or in a neighboring circuit. An *inductor* is an electrical circuit component that exhibits the properties of inductance. If a coil of wire creates self-induction, creating counterelectromotive force (CEMF), then it is referred to as an inductor. A circuit has inductance when magnetic induction occurs. *Inductive reactance* is the opposition of inductance to alternating current in a circuit. Inductive reactance (X_L) is the effect present which inhibits alternating current flow in a circuit. Bends or kinks in a cable will cause more inductance in the circuit and will likely result in problems with the signal.

Impedance (Z) is the total opposition to alternating current in a twisted pair. Impedance is the total effect of series resistance (R), series inductive reactance (X_L), and shunt capacitance (X_C) of a twisted pair. Impedance of a twisted pair is constant regardless of length but varies with frequency due to structural non-uniformities in the twisted pair. Impedance is a unit of measure expressed in ohms. Across the voiceband, it varies from about 2,300 ohms at 100 hertz to approximately 400 ohms at 3,000 hertz, with a nominal value of 600 ohms. At frequencies from 3 to 300 kilohertz, impedance can vary. At frequencies above 300 kilohertz, impedance is typically 100 ohms ($\pm15\ \Omega$). The ANSI/TIA-568.2-D requirement for the impedance of UTP cables at frequencies above one megahertz is 100 ohms ($\pm15\ \Omega$). **See Figure 4-8.**

A balanced transmission line has an equal level of impedance from either side of the line to ground, allowing for the cancellation of common mode signals that are typically the result of stray electromagnetic interference (EMI). An unbalanced transmission line is only grounded on one side, with resistance and inductance on the opposite side. Unbalanced lines are unable to cancel out EMI because current flow on the ground side is flowing through zero ohms (impedance of an ideal ground plane) and unable to develop the counteracting magnetic field required to cancel out energy radiating from the resistance and inductance on the opposite side of the line. *Conductance* (G) is the ability of an electrical circuit or component to pass (conduct) current. Conductance can be a source of EMI in a transmission line.

There are a number of factors to consider when installing cable that can affect the electrical characteristics of a data

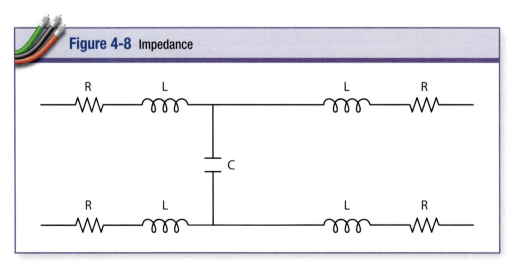

Figure 4-8 Impedance

Figure 4-8. *Impedance for a given length of twisted pair is constant, but varies with frequency.*

Tech Fact

Tip and Ring Conductors

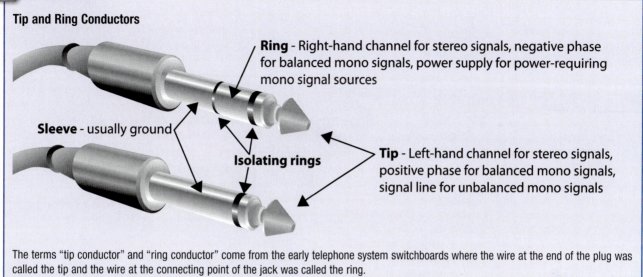

Ring - Right-hand channel for stereo signals, negative phase for balanced mono signals, power supply for power-requiring mono signal sources

Sleeve - usually ground

Isolating rings

Tip - Left-hand channel for stereo signals, positive phase for balanced mono signals, signal line for unbalanced mono signals

The terms "tip conductor" and "ring conductor" come from the early telephone system switchboards where the wire at the end of the plug was called the tip and the wire at the connecting point of the jack was called the ring.

circuit. Most are common mistakes and can be prevented. Common mistakes include:

- Excess pulling tension (over 25 foot pounds of pressure)
- Bend radius exceeding the maximum
- Cinching cables too tightly with cable ties
- Excessive cable jacket removal
- Cable combing or dressing

These common mistakes are the primary reason the ANSI/TIA standards require specific tests for structured cabling channel parameters.

STRUCTURED CABLING PARAMETERS

Unshielded twisted-pair cabling performance is dependent on several different parameters. Performance criteria such as attenuation; crosstalk; attenuation-to-crosstalk ratio (ACR); near-end crosstalk (NEXT); power sum NEXT (PSNEXT); attenuation-to-crosstalk ratio, far-end (ACRF); power sum attenuation-to-crosstalk ratio, far-end (PSACRF); delay and delay skew; impedance; return loss; and alien crosstalk (for 10 Gigabit Ethernet

over copper) are used to determine the category of compliance for the system.

Decibels

A *decibel (dB)* is a unit of measure used to express a ratio between two voltages or powers. Most parameters used for defining the performance of cables and related components are specified in decibels. Because the decibel specifies a ratio rather than an absolute value of voltage or power, it is convenient to use. Decibels reflect the use of a logarithmic scale, which means that a wide range of voltage and power levels may be covered using a small range of numbers to express those ratios in decibels. Two different ratios, the voltage ratio and the power ratio, are used to express the relationship of decibels to voltage and power. The voltage ratio is calculated by applying the following formula:

$$NdB = 20 \times log (V_0 \div V_1)$$

where:

NdB = gain or loss (in decibels)
20 = constant
log = logarithm function
V_0 = voltage out (in volts)
V_1 = voltage in (in volts)

Example: What is the gain or loss of an audio attenuator with voltages of three volts out and 1.5 volts in?

$$NdB = 20 \times log\ (V_0 \div V_1)$$
$$= 20 \times log\ (3\ V \div 1.5\ V)$$
$$= 20 \times log\ (2)$$
$$= 20 \times 0.3$$
$$= 6\ \textbf{dB}$$

This equates to a voltage ratio of 2:1; therefore, an audio attenuator with a loss of six decibels would deliver half the voltage at the output as it would at the input.

The power ratio is calculated by applying the following formula:

$$NdB = 10 \times log\ (P_0 \div P_1)$$

where:

NdB = gain or loss (in decibels)
10 = constant
log = logarithm function
P_0 = power out (in watts)
P_1 = power in (in watts)

Example: What is the gain or loss of an amplifier with power of eight watts out and two watts in?

$$NdB = 10 \times log\ (P_0 \div P_1)$$
$$= 10 \times log\ (8\ W \div 2\ W)$$
$$= 10 \times log\ (4)$$
$$= 10 \times 0.6$$
$$= 6\ \textbf{dB}$$

For every three-decibel gain, power is doubled. For every three-decibel loss, power is halved. Therefore, an amplifier with a gain of six decibels would deliver four times the power at the output as it would at the input.

Balance

Transmissions over a twisted pair rely on signal symmetry or *balance* between the two conductors of that pair. Maintaining the proper balance ensures that cabling systems and components do not emit unwanted electromagnetic radiation and are not susceptible to electrical noise. The symmetry between pairs of a cable can be distorted by improper installation techniques such as exceeding the cable's bend radius and/or exceeding the pull tension of the cable during the install. Component balance requirements are specified for category 6/Class E cabling. Component and cabling balance requirements are specified for category 6A/Class EA and higher grades of cabling.

Insertion Loss (Attenuation)

Insertion loss is the weakening of a signal as it travels down the length of a cable. Insertion loss is a measure of how much a signal is reduced in amplitude or strength as it is carried over a twisted pair. Insertion loss typically results from the signal loss that occurs from the insertion of cabling or a component between a transmitter and a receiver. Insertion loss is expressed in decibels per unit length, such as dB/km. Low insertion loss is desirable because the higher the insertion loss, the weaker the signal. Insertion loss increases as the frequency of the signal is increased. (In testing, lower numbers are better.) **See Figure 4-9.**

Crosstalk

Crosstalk is any unwanted reception of signals induced on a communication line from another communication line or from an outside source. Crosstalk causes problems in data lines, and crosstalk problems are amplified as the frequency is increased. As data moves at faster speeds, crosstalk problems will only increase. Crosstalk is caused by conductor placement, shielding, and transmission techniques. (In testing, lower numbers are better.)

Near-End Crosstalk

Near-end crosstalk (NEXT) is the undesired coupling of signal energy from a transmitting conductor pair into a receiving conductor pair nearest the point of transmission. NEXT is a basic measurement of how well adjacent pairs in a cable are electrically isolated from one another. Near-end crosstalk is expressed in decibels and is an important measurement for systems that utilize one pair for

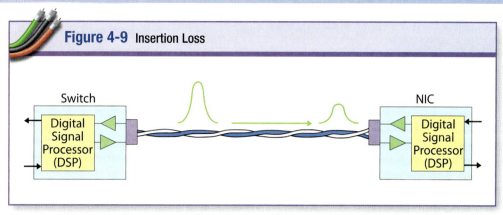

Figure 4-9. *Attenuation (also called loss or insertion loss) is the weakening of a signal as it travels down the cable.*

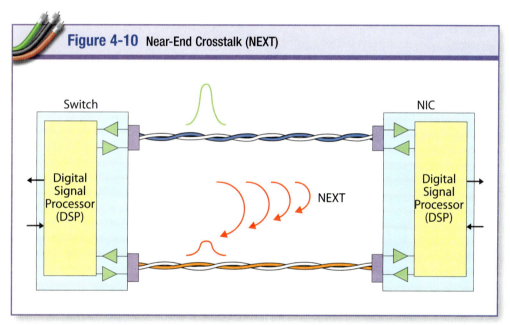

Figure 4-10. *NEXT causes undesired leakage from the transmit pair onto the near-end receive pair of the cabling.*

transmitting a signal and an adjacent pair for receiving a signal. The higher the measured value, the better the crosstalk isolation is between the pairs. NEXT degrades as the frequency of the signal is increased. (In testing, higher numbers are better.) **See Figure 4-10.**

Power Sum Near-End Crosstalk

Power sum near-end crosstalk (PSNEXT) is the undesired coupling of signal energy from a number of simultaneously transmitting pairs onto a receiving pair. **See Figure 4-11.** PSNEXT is the sum of the NEXT power from all other pairs in the cable at the near end of a cable, link, or channel. PSNEXT is an important measurement for systems that utilize more than one pair for transmitting signals while using multiple adjacent pairs for receiving signals, such as Gigabit Ethernet. (In testing, higher numbers are better.) **See Figure 4-12.**

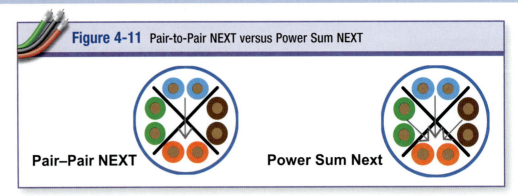

Figure 4-11 Pair-to-Pair NEXT versus Power Sum NEXT

Pair–Pair NEXT **Power Sum Next**

Figure 4-11. *Near-end crosstalk (NEXT) involves the undesired coupling of signal energy from a transmitting conductor pair (pair-to-pair NEXT) or a number of trasmitting conductor pairs (power sum NEXT) into a receiveing pair.* Courtesy of Ideal, Inc.

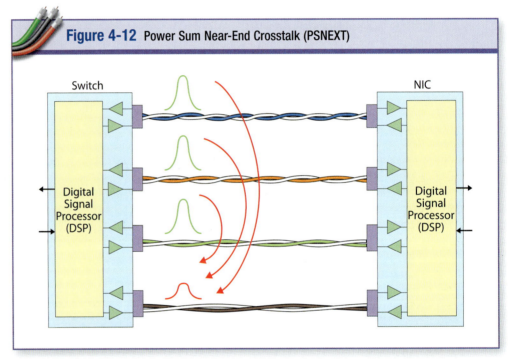

Figure 4-12 Power Sum Near-End Crosstalk (PSNEXT)

Switch

NIC

Digital Signal Processor (DSP)

Digital Signal Processor (DSP)

Figure 4-12. *Power sum NEXT (PSNEXT) is the sum of the NEXT power from all other pairs in the cable.*

Attenuation-to-Crosstalk Ratio

Attenuation-to-crosstalk ratio (ACR) is the ratio of the power of the signal received (attenuated by the media) over the power of the NEXT from the local transmitter. **See Figure 4-13**. Attenuation-to-crosstalk ratio is the difference between attenuation and crosstalk measured in decibels at a given frequency. This difference is critical in network transmissions to ensure that the signal sent down the twisted-pair cable is stronger at the receiving end of the cable than any interference signals (crosstalk) from other cable pairs. ACR can be explained as the ratio of the worst-case attenuation of pairs in a cable to the worst-case pair-to-pair NEXT. The ACR defines the difference between connector attenuation and NEXT for systems that include connectors and interfaces in their ratings. (In testing, higher numbers are better.) **See Figure 4-14**.

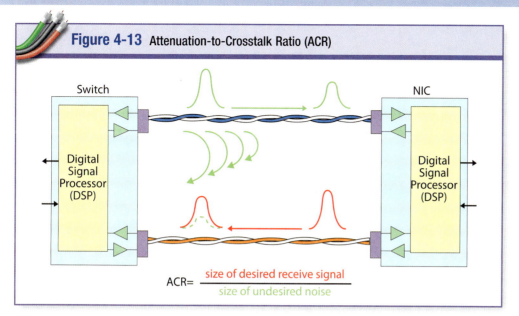

Figure 4-13 Attenuation-to-Crosstalk Ratio (ACR)

$$ACR = \frac{\text{size of desired receive signal}}{\text{size of undesired noise}}$$

Figure 4-13. ACR is equal to the size of the desired receive signal divided by the size of undesired noise.

Power Sum Attenuation-to-Crosstalk Ratio

Power sum attenuation-to-crosstalk ratio (PSACR) is a measure of the quality of a transmission channel in a wiring technology with multiple wire pairs. PSACR is the difference between PSNEXT and insertion loss (attenuation). Category 5e cables should have a PSACR of 10.3 decibels at 100 megahertz. Category 6 cables should have a PSACR of 22.5 decibels at 100 megahertz. (In testing, higher numbers are better.)

Far-End Crosstalk

Far-end crosstalk (FEXT) is a type of crosstalk, expressed in decibels, that occurs when signals on one twisted pair are coupled to another twisted pair within the same cable sheath as they arrive at the far end of a multi-pair cable system. The stronger signal distorts the weaker one. Noise caused by FEXT can be canceled out, decreasing the noise level.

Attenuation-to-Crosstalk Ratio, Far-End

Attenuation-to-crosstalk ratio, far-end (ACRF) (previously known as *equal level far-end crosstalk,* or *ELFEXT*) is pair-to-pair far-end crosstalk loss which is the undesired signal coupling between adjacent pairs at the far end (the opposite end of the transmit end) of a cable or component. ACRF is calculated by subtracting the measured insertion loss from the measured far-end crosstalk loss. It equals a normalized value that can

Tech Fact

While FEXT does not cause as many problems as NEXT does, modern high-speed systems have been designed to eliminate NEXT so well that the cables used in these systems require FEXT ratings as the main test parameters for compliance.

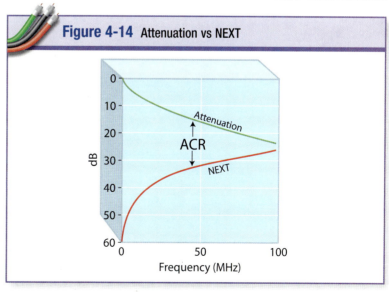

Figure 4-14 Attenuation vs NEXT

Figure 4-14. A graphic relationship of attenuation and NEXT plotted over the frequency domain shows convergence as the frequency increases.

be used to compare cable and cabling performance independent of length. Poor ACRF levels result in increased bit error rates and/or undeliverable data packets. Note that NEXT loss margin alone is not sufficient to ensure proper ACRF performance. (In testing, higher numbers are better.) **See Figure 4-15**.

Power Sum Attenuation-to-Crosstalk Ratio, Far-End

Power sum attenuation-to-crosstalk ratio, far-end (*PSACRF*) (previously known as *power sum equal level far-end crosstalk*, or *PSELFEXT*) is the undesired coupling of signal energy from a number of simultaneously-transmitting pairs into a receiving pair. PSACRF is calculated at the opposite end of the cable from where the transmitter is located. There are four PSACRF results for each cable end. PSACRF is an especially important measurement for systems that utilize more than one pair for transmitting signals while using multiple adjacent pairs, or

the same pairs, for receiving signals. PSACRF is the sum of the attenuation-to-crosstalk ratio, far-end (ACRF) power from all other conductor pairs in the cable. (In testing, higher numbers are better.) **See Figure 4-16**.

Delay and Delay Skew

Delay is the amount of time needed for a signal to travel over a twisted pair, measured in nanoseconds. Because each of the four pairs in a cable have a different number of twists per inch, the electrical length of each pair will be different. The amount of time it takes for a signal to travel from the transmitting end to the receiving end of a pair (the delay) is different for each of the pairs within the cable.

Delay skew is the disparity in the delay between any two pairs within the same cable. Delay skew is the difference between the amount of delay for the fastest (longest twist length) and slowest (shortest twist length) pair. **See Figure 4-17**.

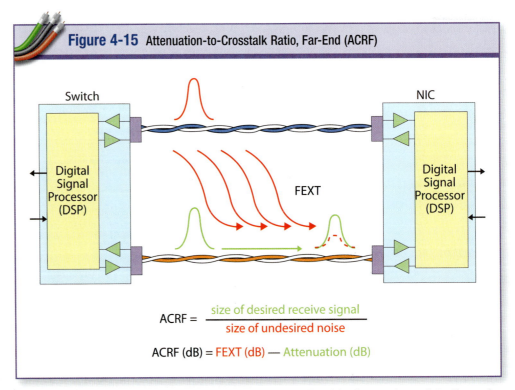

Figure 4-15 Attenuation-to-Crosstalk Ratio, Far-End (ACRF)

$$ACRF = \frac{\text{size of desired receive signal}}{\text{size of undesired noise}}$$

$$ACRF (dB) = FEXT (dB) - Attenuation (dB)$$

Figure 4-15. *ACRF is the ratio of the desired receive signal in the receive pair to the undesired FEXT noise. ACRF is the equivalent of ACR for far-end coupling.*

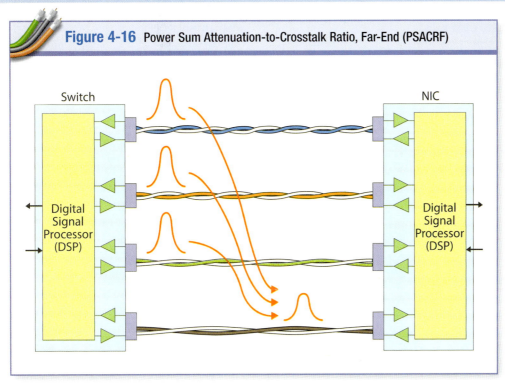

Figure 4-16 Power Sum Attenuation-to-Crosstalk Ratio, Far-End (PSACRF)

Figure 4-16. PSACRF is the sum of the ACRF power from all other pairs in the cable.

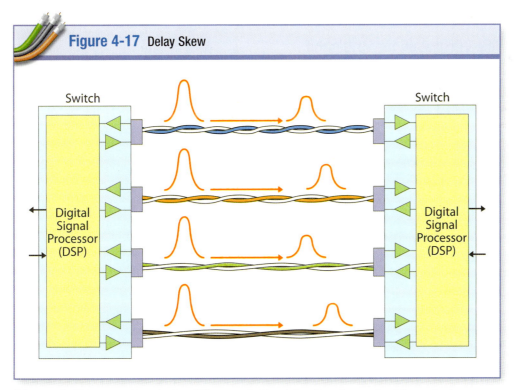

Figure 4-17 Delay Skew

Figure 4-17. Delay skew is the difference in the delay time for each of the pairs in a 4-pair cable.

Return Loss

Return loss (RL) is the ratio of the transmitted signal to the reflected signal of the cabling. Return loss is a measurement of the amount of signal energy that is reflected back to a transmitter as it travels down a twisted pair while encountering impedance discontinuities. The measurement is expressed in decibels and is a ratio of the amplitude of the transmitted signal with respect to the amplitude of the reflected signal. Impedance discontinuities occur at connections where cable is terminated to plugs or jacks and within the plug/jack connection. An impedance discontinuity can also occur if a cable is bent too much, kinked, or otherwise damaged. When a transmitted signal pulse hits one of these structural discontinuities, return loss occurs. Too much RL added to the PSNEXT and PSFEXT adds to the noise, resulting in increased bit error rate, lower signal-to-noise ratio, less network operating margin, and more downtime. (In testing, a lower absolute value is better.) **See Figure 4-18.**

Alien Crosstalk for 10 Gigabit Ethernet (Copper)

Alien crosstalk (AXT) is an unwanted signal coupling from one balanced twisted-pair component, channel, or permanent link to another. Because of the high frequencies used in 10 Gigabit Ethernet over copper transmissions, these systems may need to be tested for the presence of alien crosstalk. The term "alien" is used to describe this form of crosstalk because it occurs between different cables in a group or bundle, rather than between individual conductors or pairs within a single cable sheath. For example, alien crosstalk can occur when six (disturbing) cables surround one (victim) cable, with the alien crosstalk present in the victim cable. This type of crosstalk is only caused by balanced signal coupling between pairs that have the same amount of twist. Common mode noise sources such as motors, fluorescent lights, and transformers are not a factor. **See Figure 4-19.**

Alien crosstalk field measurements for 10 Gigabit Ethernet on copper UTP may require two additional tests not performed on category 6 cable: power sum attenuation-to-alien-crosstalk ratio, far-end (PSAACR-F), and power sum alien near-end crosstalk (PSANEXT) testing.

Power Sum Attenuation-to-Alien-Crosstalk Ratio, Far-End

Power sum attenuation-to-alien-crosstalk ratio, far-end (PSAACR-F) is the measurement of the alien crosstalk ratio, corrected for the length of the link when the disturber link is stimulated from the near end of the link and the victim link is measured from the far end of the link. This test is performed in both directions; therefore, an "NE" or "FE" will be appended

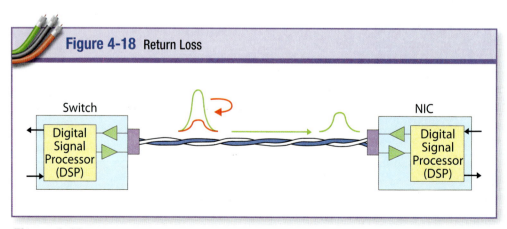

Figure 4-18 Return Loss

Figure 4-18. *Return loss (also called echo response) is the ratio of the transmit signal to the reflected signal of the cabling.*

to the PSAACR-F on the reports to indicate from which direction the test data applies.

Power Sum Alien Near-End Crosstalk

Power sum alien near-end crosstalk (PSANEXT) is the measurement of alien crosstalk when the disturber link is stimulated at the same side of the link where the victim link is measured. This test is run from both sides of the link and results in two sets of data; therefore, an "NE" or "FE" will be appended to the PSANEXT on the reports to indicate from which direction the test data applies.

UTP CABLING CATEGORIES

Both the ISO/IEC and TIA Standards organizations have defined generic cabling systems suitable for medium and large offices. Details can be found in the ISO/IEC IS 11801, *Standard for Customer Premises Cabling,* and in ANSI/TIA-568.1-D.

ISO/IEC 11801, ANSI/TIA-568.1-D, and the European version, EN 50173-1, are all key standards for cabling installation. All of these cover similar areas, but use different approaches to achieve conformity. ISO/IEC 11801 is a global standard that has evolved to meet the needs of all geographic areas. As a result, some of its requirements are very broad.

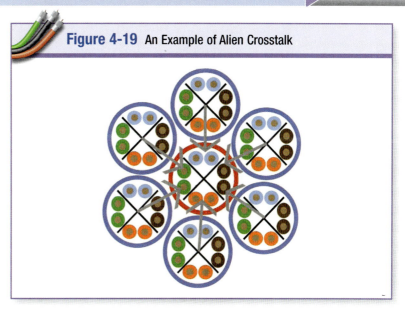

Figure 4-19 An Example of Alien Crosstalk

Figure 4-19. Alien crosstalk sometimes involves multiple disturber cables affecting a single cable in a bundle. Courtesy of Ideal, Inc.

ANSI/TIA-568.1-D and ISO/IEC 11801 specify several cabling categories. The first two categories are suited only to voice and data communications up to four megabits per second and are seldom used in data networking applications. While both TIA and ISO/IEC refer to components in terms of "categories," TIA also uses "categories" for end-to-end channel performance, while the ISO/IEC standard refers to the end-to-end performance in terms of "Classes." **See Figure 4-20.**

Figure 4-20 TIA and ISO Equivalent Classifications

TIA and ISO Equivalent Classifications				
Frequency Bandwidth	TIA (Components)	TIA (Cabling)	ISO (Components)	ISO (Cabling)
1 - 100 MHz	Category 5e	Category 5e	Category 5e	Class D
1 - 250 MHz	Category 6	Category 6	Category 6	Class E
1 - 500 MHz	Category 6A	Category 6A	Category 6A	Class E_A
1 - 600 MHz	not supported	not supported	Category 7	Class F
1 - 1,000 MHz	not supported	not supported	Category 7_A	Class F_A
1 - 2,000 MHz	Category 8	Category 8	Category 8	Class I

Figure 4-20. TIA classifies cables using Category, while ISO uses Classes for cables.

CATEGORY DESCRIPTIONS

Specifications for UTP cabling standards define performance requirements for cables, connectors, basic links, and channels. UTP cabling standards define requirements for:

- Category 3 (carrier frequency up to 16 megahertz)
- Category 4 (carrier frequency up to 20 megahertz)
- Category 5 (carrier frequency up to 100 megahertz)
- Category 5e (carrier frequency up to 100 megahertz, but with more stringent requirements than category 5)
- Category 6 (carrier frequency up to 250 megahertz)
- Category 6e (never ratified in the standards; carrier frequency up to 250 megahertz, test as category 6)
- Category 6A (carrier frequency up to 500 megahertz)
- Category 8 (carrier frequency up to 2,000 megahertz)

As the category number increases, overall performance increases, attenuation decreases, and crosstalk performance increases.

Category 3 cabling systems were primarily used for voice applications while categories 4 and 5 were once used for data applications, but all three have been removed from current standards. Most UTP cabling for new installations is compliant with category 5e, 6, or 6A requirements. **See Figure 4-21.**

Figure 4-21 TIA and ISO Standards References

TIA Cable Category and Ratification Date	
Category 5e	ANSI/TIA/EIA-568-B.2, *Balanced Twisted-Pair Telecommunications Cabling and Components Standard* Part 2: *Balanced Twisted Pair Cabling Components,* 2001
Category 6	ANSI/TIA/EIA-568-B.2-1, *Commercial Building Telecommunications Standard,* Part 2 Addendum 1: Transmission Performance Specifications for 4 Pair 100 ohm Category 6 cabling, 2002
Category 6A	ANSI/TIA/EIA-568-B.2-10, *Commercial Building Telecommunications Standard,* Part 2 Addendum 10: Transmission Performance Specifications for 4 Pair 100 ohm Augmented Category 6 Cabling, 2008
Category 8	ANSI/TIA-568-C.2 *Commercial Building Telecommunications Standard.* *Balanced Twisted-Pair Telecommunications Cabling and Components Standard,* Addendum 1: Specifications for 100 Ω Category 8 Cabling TIA-568-C.2-1 (Addendum to TIA-568-C.2) (July 2016)
ISO Class and Ratification Date	
Class D	ISO/IEC 11801, 2nd Ed., *Information technology - Generic Cabling for Customer Premises,* 2002
Class E	ISO/IEC 11801, 2nd Ed., *Information technology - Generic Cabling for Customer Premises,* 2002
Class E$_A$	Amendment 1 to ISO/IEC 11801, 2nd Ed., *Information technology - Generic Cabling for Customer Premises,*
Class F	ISO/IEC 11801, 2nd Ed., *Information technology - Generic Cabling for Customer Premises,* 2002
Class F$_A$	Amendment 1 and 2 to ISO/IEC 11801, 2nd Ed., *Information technology - Generic Cabling for Customer Premises,*
Class I	link/channel up to between 1,600 and 2,000 MHz using Category 8.1 cable/connectors ISO/IEC 11801-1 in 2017

Figure 4-21. New cable categories and classes have been developed through the years.

Category 5e

This cable consists of 100-ohm twisted-pair copper cable that meets or exceeds specifications in ANSI/TIA-568.2-D, and ISO/IEC IS 11801, for frequencies up to 100 megahertz. Its performance requirements are more severe than those for category 5.

Category 5e (Enhanced Category 5) is an upgrade to category 5 specifications that was targeted as minimally compliant support of Gigabit Ethernet (1000Base-T). The maximum frequency specified for categories 5 and 5e is 100 megahertz. One gigabit per second signaling is accomplished via PAM5 encoding scheme transmitted over all cable pairs.

ISO/IEC refers to the end-to-end category 5e channel as a Class D channel.

Category 6

This cable consists of 100-ohm twisted-pair copper cable that meets or exceeds specifications in ANSI/TIA-568.2-D and ISO/IEC 11801 at frequencies up to 250 megahertz. Its performance requirements are more severe than those for category 5e.

Category 6 cabling was designed with a significant improvement in bandwidth, which is over double the bandwidth of category 5e for support of Gigabit Ethernet (1000Base-T) over the maximum frequency specified of 250 megahertz. Category 6 cabling is also recommended if mid-span power over Ethernet (PoE) is required to be supported as an application since the additional connections introduced by the mid-span equipment may have a detrimental effect on the crosstalk and return loss performance of the end-to-end channel.

ISO/IEC refers to the end-to-end category 6 channel as a Class E channel.

Category 6e

This cable consists of an enhanced or "extended" bandwidth version of category 6 cable. Basically, category 6e cable is being offered by the cabling industry as a more "advanced" cable than a category 6 cable. Manufacturers claim that frequencies for their category 6e offerings may extend from 350 megahertz to 550 megahertz. Category 6e cable is offered by cabling manufacturers to claims of increased headroom allowing higher throughput for broadband video, Gigabit Ethernet, 155 Mb/s ATM, 100 Mb/s TP-PMD/CDDI, and fast Ethernet.

Note that there is no category 6e standard, and this cable falls under the same installation testing parameters and specifications as category 6 cable (250 MHz).

Category 6A

This cable consists of 100-ohm twisted-pair copper cable that is designed to meet or exceed the requirements of 10 Gigabit Ethernet (10GBase-T). It extends the cabling bandwidth beyond category 6 by specifying the frequency range out to 500 megahertz, and includes the alien crosstalk specifications that are vital for the support of 10 Gigabit Ethernet.

ISO/IEC 11801 amendment 2 includes the Class E_A specification, offering better performance than category 6A for certain parameters. TIA Standards for category 6A were ratified on February 8, 2008.

Class F

Class F cable consists of 100-ohm shielded twisted-pair cabling that meets or exceeds specifications in only ISO/IEC 11801 and CENELEC EN50173-1 as Class F cable. The equivalent category 7 cable has never been ratified in any of the TIA-568 standards.

Class F cable and connecting hardware products have transmission characteristics specified to 600 megahertz and require individually shielded twisted-pair cables. Class F also requires terminating with either a switched RJ45 or a non-RJ45 style of connector, neither of which have been standardized by ANSI/TIA.

Although Class F has been standardized internationally since 2002, because of the requirement for individual pair shielding and choice of connectors, it has not found worldwide acceptance and is not widely used.

Category 8

This cable was ratified by the TR43 working group under ANSI/TIA-568-C.2-1 in July 2016. It is defined up to 2,000 megahertz and only for distances up to 30 meters (98 ft) depending on the patch cords used. Category 8 is designed only for data centers where distances between switches and servers is short. It is not intended for general office cabling.

Class F$_A$

The significant enhancement in Class F$_A$ specifications is the extension of the frequency bandwidth of characterization from 600 megahertz to 1,000 megahertz. This allows Class F$_A$ cabling to be capable of supporting all channels of broadband video that operate up to 862 megahertz (as in a CATV application). Augmented Class F was approved on February 8, 2008 at the same time category 6A was ratified.

Although the TIA is not actively developing a standard for Class F or Class F$_A$ at this time, it is acceptable to specify Class F cabling in the USA and other markets. The rationale for this is that, in addition to being recognized by BICSI, NEMA, IEEE, and other standards organizations, Class F is simply a superset of TIA category 6A requirements. Field test

requirements and adapters for Class F cabling qualification have been commercially available since 2002.

ISO ratified the equivalent with two options:

- Class I channel (category 8.1 cable): minimum cable design U/FTP or F/UTP, fully backward compatible and interoperable with Class EA (category 6A) using 8P8C connectors
- Class II channel (category 8.2 cable): F/FTP or S/FTP minimum, interoperable with Class F$_A$ (category 7A) using TERA or GG45 connectors. **See Figure 4-22.**

APPLICATIONS SUPPORT

Table 5 in TIA-568.0-D-1 summarizes cabling types capable of supporting commonly specified applications over 100 meters, using twisted 4-pair connector topologies. **See Figure 4-23.**

FIELD TESTING OF UTP CABLING SYSTEMS

In October 1995, the TIA published Telecommunications Systems Bulletin (TSB)-67, *Transmission Performance Specifications for Field Testing of Unshielded Twisted Pair Cabling Systems.* This document specified the electrical characteristics of field tes-

---

 Tech Fact

Cabling for 10GBase-T

The IEEE developed the 10GBase-T Standard with the intent of targeting the installed base of category 6/Class E cabling. However, although category 6/Class E cabling is the minimum requirement in the 10GBase-T Standard, achievable distances over category 6/Class E cabling are highly dependent on the alien crosstalk environment as well as the cabling performance at frequencies not specified for category 6/Class E. Shielded category 6/Class E cabling may exhibit improved alien crosstalk, but without a normative specification and without the necessary standards specifications for higher frequency performance.

Guidelines for the verification of 10GBase-T requirements have been developed by EIA/TIA (TSB-155) and ISO/IEC (TR 24750), but these do not provide improved specifications for category 6/Class E and are not intended for new installations. Achieving useful distances over UTP is likely to require multiple mitigation steps ranging from the unbundling of cables to component replacement, since the category 6/Class E cabling standards fall considerably short of the 10GBase-T requirements and do not specify alien crosstalk.

Figure 4-22 Comparing Standards

Parameter	Cat 5 (1995)	Cat 5 (1999)	Cat 5e	Cat 6
Max. Frequency (MHz)	100	100	100	250
Attenuation (dB)	Testing Required	Testing Required	Testing Required	Testing Required
Pr-Pr NEXT (dB)	Testing Required	Testing Required	3 dB > Cat 5	10 dB > Cat 5e
PSNEXT(dB)	Only > 4pr	Only > 4pr	Includes 4 pr.	10 dB > Cat 5e
ELFEXT(dB)	No Testing Required	Testing Required	Testing Required	6 dB > Cat 5e
PSELFEXT(dB)	No Testing Required	Testing Required	Testing Required	6 dB > Cat 5e
Delay (ns)	Only for Cable	Testing Required	Testing Required	Testing Required
Delay Skew (ns)	No Testing Required	Testing Required	Testing Required	Testing Required
Return Loss (dB)	No Testing Required	Testing Required	2 dB > Cat 5	2dB > Cat 5e

Figure 4-22. Category 5e offers a three-decibel improvement in NEXT and a two-decibel improvement in RL over category 5. Category 6 has a 10-decibel improvement in NEXT, a six-decibel improvement in FEXT, and a two-decibel improvement in RL over category 5e, as well as an extension in the frequency range over which most parameters must be tested from 100 to 250 megahertz.

ters, test methods, and minimum transmission requirements for UTP cabling.

TSB-67 applied only to 4-pair category 3, category 4, and category 5 UTP cabling. Addendums to the TIA/EIA-568-A Standard incorporated TSB-67 into the 568-A Standard and provided additional specifications for the measurement of category 5e systems. The contents of these documents were subse-

Figure 4-23 Twisted-Pair Applications

Application	Category 5e Class D	Category 6 Class E	Category 6A Class E_A	Class F	Class F_A	Category 8
10BaseT	x	x	x	x	x	x
100BaseT4	x	x	x	x	x	x
155 Mb/s ATM	x	x	x	x	x	x
1000BaseT	x	x	x	x	x	x
TIA/EIA-854		x	x	x	x	x
2.5GbaseT		x	x	x	x	x
5GbaseT			x	x	x	x
10GbaseT			x	x	x	x
25GbaseT			x	x	x	x
40GbaseT				x	x	x
Broadband CATV				x	x	

Figure 4-23. A partial applications chart shows the different categories and classes of cabling.

quently incorporated into ANSI/TIA/EIA-568-B.1. ANSI/TIA/EIA-568-B.2-1 provided additional specifications for the measurement of category 6 systems. ANSI/TIA-568-C.2 incorporated performance specifications for category 6 and category 6A cables. ANSI/TIA-568-C.2.2 added alternative test methodology for category 6A patch cords. ANSI/TIA-568-C.2-1 incorporated specifications for category 8 cabling. Two different test configurations cited by the standards include permanent link and channel.

Permanent Link Test Configurations

A *permanent link test configuration* is a test configuration used by data telecommunications designers and users to verify the performance of permanently installed cabling. A permanent link consists of up to 90 meters (295 ft) of horizontal cabling, a connection at each end, up to two meters (6.6 ft) of test equipment cord from the main unit of the field tester to the local connection, and up to two meters (6.6 ft) of test equipment cord from the remote connection to the remote unit of the field tester. The permanent link has one connection at each end of the link. **See Figure 4-24.** While test cord connections to the horizontal cabling at each end of the permanent link are included in the permanent link test configuration, the test cord connections to the field testers at each end of the permanent link are not included in the permanent link configuration. **See Figure 4-25.**

Channel Test Configurations

A *channel test configuration* is a test configuration used by data telecommunications system designers and users to verify the performance of the overall channel. The channel includes up to 90 meters (295 ft) of horizontal cable, a work area equipment cord, a telecommunications outlet, an optional consolidation connection close to the work area, and two cross-connect connections in the telecommunications room. Per ANSI/TIA-568.2-D, the total length of equipment cords, patch cords, and jumpers shall not exceed 10 meters (33 ft). **See Figure 4-26.** The connections to the equipment at each end of the channel are not included in the channel configuration. **See Figure 4-27.**

Figure 4-24 Permanent Link Testing

Figure 4-24. As per the Standards, permanent links must be tested from both ends.

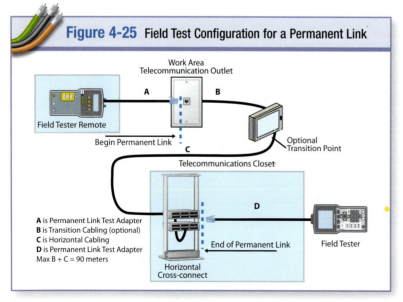

Figure 4-25 Field Test Configuration for a Permanent Link

Figure 4-25. Permanent links must be field tested to ensure optimal performance.

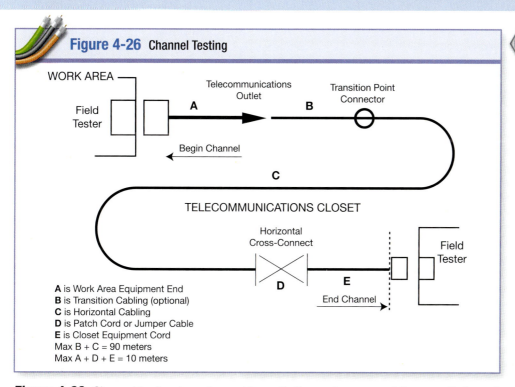

Figure 4-26. *Channel testing is performed to verify the performance of the overall channel.*

Tech Fact

A phrase that is used by installers in the field to help remember the points of a permanent link test configuration is "from rack to jack" which describes the beginning and ending test location points.

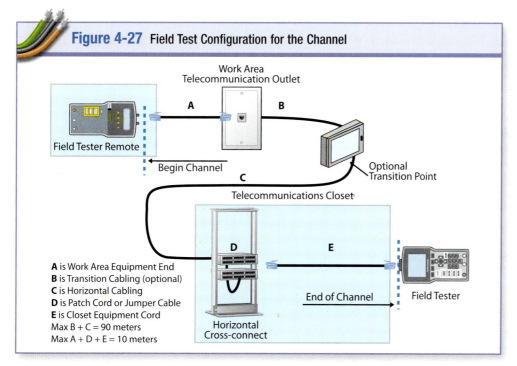

Figure 4-27. *Channel testing includes testing the patch cables that will be used as part of the system.*

Figure 4-28 Common Wiring Mistakes

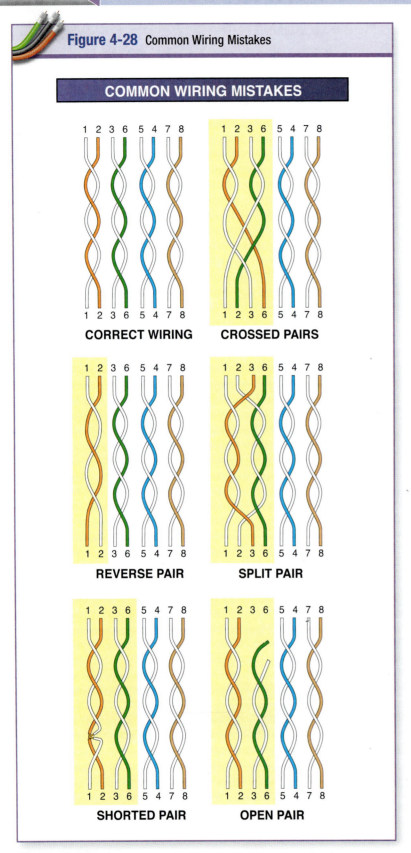

Figure 4-28. *Common wiring mistakes include crossed pairs, reverse pairs, split pairs, shorted pairs, and open pairs.*

Wire Map Tests

A *wire map test* is a basic test for testing for open or short circuits, crossed pairs, reversed wires, and split pairs. Wire map tests are performed on all voice-data-video (VDV) cables after the cable is installed. Wire map tests are used to test 4-pair cables, 25-pair cables, and all other cables. A wire map test is used to ensure that a cable is properly terminated to the correct terminals and to make certain that no wires are left unconnected or connected to the wrong terminal. A wire map tester indicates when a cable passes or fails a given test.

When a cable fails a test, understanding the fault will allow for easy correction of the problem and a pass on the re-test. Although there can be any number of wiring mistakes and problems, there are several typical ones. **See Figure 4-28.**

Length Tests

A *length test* is a test that measures the length of the link/channel on the conductor pair having the shortest delay. The maximum physical length allowed for a permanent link is 90 meters. The maximum physical length allowed for a channel is 100 meters, including equipment cords and patch cords.

Nominal Velocity of Propagation

The *nominal velocity of propagation (NVP)* is the approximate speed at which a signal moves through a cable. The NVP is expressed as a percentage of the speed of light in a vacuum. Some cable testers use this speed, along with the time it takes for a signal to return to a testing device. Pass/fail criteria are based on these maximum values plus the NVP uncertainty of ±10%. Calibration of NVP for the particular cable under test is critical to obtaining accurate length measurements.

Performance requirements vary for category 5, 5e, 6, and 6A channels. **See Figure 4-29.**

SHIELDED CABLE

The use of shielded cable is reemerging as a viable option because of the demand

Figure 4-29 UTP Channel Performance Requirements

Parameter	Category 5e Class D	Category 6 Class E	Category 6A Class E_A	Class F	Class F_A
Frequency Range (MHz)	1 - 100	1 - 250	1 - 500	1 - 600	1 - 1,000
Insertion Loss (dB)	24	21.3 (21.7)	20.9	20.8	20.3
NEXT Loss (dB)	30.1	39.9	39.9	62.9	65
PSNEXT Loss (dB)	27.1	37.1	37.1	59.9	62
ACR (dB)	6.1	18.6	18.6	42.1	46.1
PSACR (dB)	3.1	15.8	15.8	39.1	41.7
ACRF[1] (dB)	17.4	23.3	23.3 (25.5)	44.4	47.4
PSACRF[2] (dB)	14.4	20.3	20.3 (22.5)	41.4	44.4
Return Loss (dB)	10	12	12	12	12
PSANEXT Loss (dB)	"not supported"	"not supported"	60	"not supported"	67
PSAACRF (dB)	"not supported"	"not supported"	37	"not supported"	52
TCL (dB)	"not supported"	"not supported"	20.3	20.3	20.3
ELTCTL (dB)	"not supported"	"not supported"	0.5 (0)[3]	0	0
Propagation Delay (ns)	548	548	548	548	548
Delay Skew (ns)	50	50	50	30	30

Figure 4-29. *Performance requirements vary from one cable type to another.*

for twisted-pair cables to perform at higher frequencies; this will allow for greater bandwidth. Traditionally, cables have been identified as either UTP or STP. New naming conventions are being used to better describe the types of shielded cable available.

Screened Twisted Pair

Screened twisted pair (Sc/UTP) is a cable with one or more pairs of twisted copper conductors covered with an overall metallic screen or braid over the group of four pairs to provide AXT and EMI shielding. **See Figure 4-30.**

Figure 4-30 Screened Twisted Pair

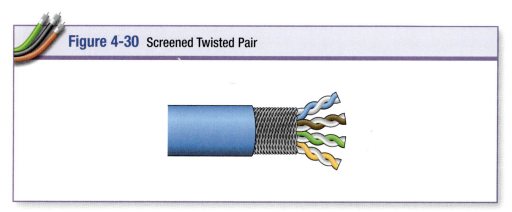

Figure 4-30. *Screened twisted pair (Sc/UTP) has an overall metallic screen or braid to shield the twisted pairs.*

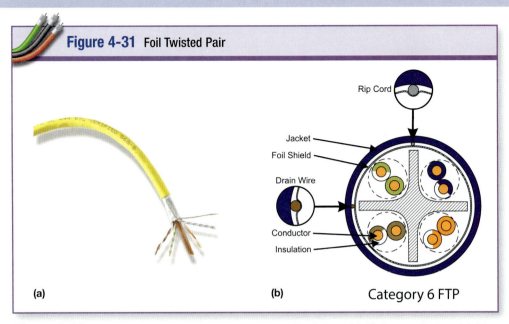

Figure 4-31 Foil Twisted Pair

Rip Cord

Jacket
Foil Shield

Drain Wire

Conductor

Insulation

(a)

(b)

Category 6 FTP

Figure 4-31. *Foil twisted pair (F/UTP) has a thin metallic foil to shield the conductors (a). The placement of this foil shield can be seen in cross-section (b).* *Courtesy of Tyco Electronics™*

Foil Twisted Pair

Foil twisted pair (F/UTP) is similar to ScTP but uses a thin metallic foil as opposed to a braid. **See Figure 4-31.**

The difference between these two cabling types is that the screen shield is more durable and easier to terminate onto a shielded jack, while the foil shield provides better coverage (no holes) and is effective at higher frequencies; however, it can easily tear if mishandled.

Pairs in Metal Foil

Pairs in metal foil (PiMF) cables have a foil shield for each pair to virtually eliminate internal crosstalk, and usually include an overall foil or screen to provide additional EMI immunity. PiMF cables are sometimes referred to as double-shielded. **See Figure 4-32.**

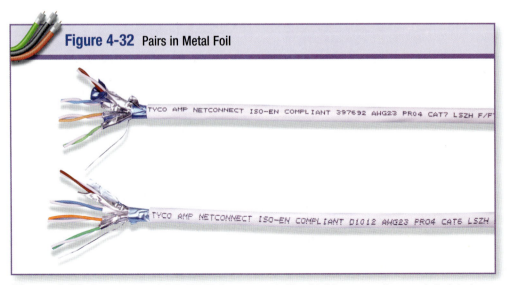

Figure 4-32 Pairs in Metal Foil

TYCO AMP NETCONNECT ISO-EN COMPLIANT 397692 AWG23 PRO4 CAT7 LSZH F/F

TYCO AMP NETCONNECT ISO-EN COMPLIANT D1012 AWG23 PRO4 CAT6 LSZH

Figure 4-32. *Pairs in metal foil (PiMF) have a metallic shield surrounding each twisted pair. Courtesy of Tyco Electronics™*

SUMMARY

Cabling performance is all about delivering data at the required rate without any errors. To deliver the required performance, the cable has to transmit a signal with little or no noise or delay. As the frequency of the signal on the cable is increased, there is a change in inductive (X_L) and capacitive (X_C) reactance of the cable. Reactance has an effect on how long the cable can be at a given frequency before signal degradation occurs. Inductive effects tested for include crosstalk (NEXT, FEXT, ACRF, and power sum). The effect of capacitance on a cable is the distortion of the signal. Capacitance slows the rise and fall time of a digital transmission signal.

Each of the parameters for cable performance is testable. Cable certifiers and qualifiers check all of the cable parameters to make sure the link will perform per the standards. Understanding what constitutes cable performance and what is being measured is required by the installer to verify the link as well as to troubleshoot problems that may arise from improper installation. Knowing what cable performance parameter is out of spec indicates where to begin to look for a solution.

REVIEW QUESTIONS

1. What is the smallest unit of information that a computer can process?
   a. Bit
   b. Byte
   c. Decibel
   d. Octet

2. Modern equipment can send data signals of up to ___?___ over UTP cabling at distances of up to 100 meters.
   a. 1 Gbp/s
   b. 2.5 Gbp/s
   c. 5 Gbp/s
   d. 10 Gbp/s

3. When energy is transferred from one circuit to another by virtue of mutual capacitance, it is known as ___?___.
   a. attenuation-to-crosstalk ratio
   b. capacitive coupling
   c. far-end crosstalk
   d. near-end crosstalk

4. ___?___ is the undesired coupling of signal energy from a transmitting conductor pair into a receiving conductor pair nearest the point of transmission.
   a. Crosstalk
   b. Insertion loss
   c. Near-end crosstalk
   d. Power sum near-end crosstalk

5. ___?___ is unwanted reception of signals induced on a communication line from another communication line or from an outside source.
   a. Crosstalk
   b. Insertion loss
   c. Near-end crosstalk
   d. Power sum near-end crosstalk

6. ___?___ is the weakening of a signal as it travels down the length of a cable.
   a. Crosstalk
   b. Insertion loss
   c. Near-end crosstalk
   d. Power sum near-end crosstalk

7. ___?___ is the total opposition to alternating current in a twisted pair.
   a. Capacitance
   b. Impedance
   c. Insertion loss
   d. Return loss

8. ___?___ is the ratio of the transmitted signal to the reflected signal of the cabling.
   a. Capacitance
   b. Capacitive reactance
   c. Insertion loss
   d. Return loss

9. Which type of UTP cable has a frequency bandwidth of 500 megahertz?
   a. Category 5e
   b. Category 6
   c. Category 6A
   d. Category 8

10. Which type of UTP cable has a frequency bandwidth of up to 2,000 megahertz?
   a. Category 5e
   b. Category 6
   c. Category 6A
   d. Category 8

Pathways and Spaces

Introduction

The structured cabling standard ANSI/TIA-569-E, *Telecommunication Pathways and Spaces,* provides guidelines for the design of horizontal, backbone, and work area cabling pathways as well as building entrance facilities, telecommunications rooms, and equipment rooms. It also gives direction on heating, ventilation and air conditioning (HVAC), firestopping, and the separation requirements for electromagnetic interference (EMI). A working knowledge of this information is vitally important to anyone who is going to install and/or service structured cabling systems.

Buildings are a large capital investment and therefore are designed to have a long life cycle. Telecommunications technology placed and used within a building is often obsolete shortly after it is installed. The building structured cabling system needs to be designed and installed in such a way to support the changes that will occur as technology changes. The cabling infrastructure must also support a multi-product, multi-vendor environment.

Objectives

- State the purpose of the ANSI/TIA-569-E standard for *Telecommunications Pathways and Spaces.*
- Define terms associated with the design and construction of telecommunications pathways and spaces.
- Identify requirements for specific pathway installations that must be met in order to be compliant with the standard.

Chapter 5

Table of Contents

PATHWAYS AND SPACES

The purpose for standardizing specific pathway and space design in construction practices in support of telecommunications media and equipment within buildings is threefold:

1. Buildings are dynamic.
2. Building telecommunications systems are dynamic.

3. More of the building systems and control can be supported over the same cabling infrastructure.

In many instances, the building is a tenant-occupied lease space and the building design should reflect the dynamic nature of the various occupants moving in and out of the space after the building has been provisioned. **See Figure 5-1.**

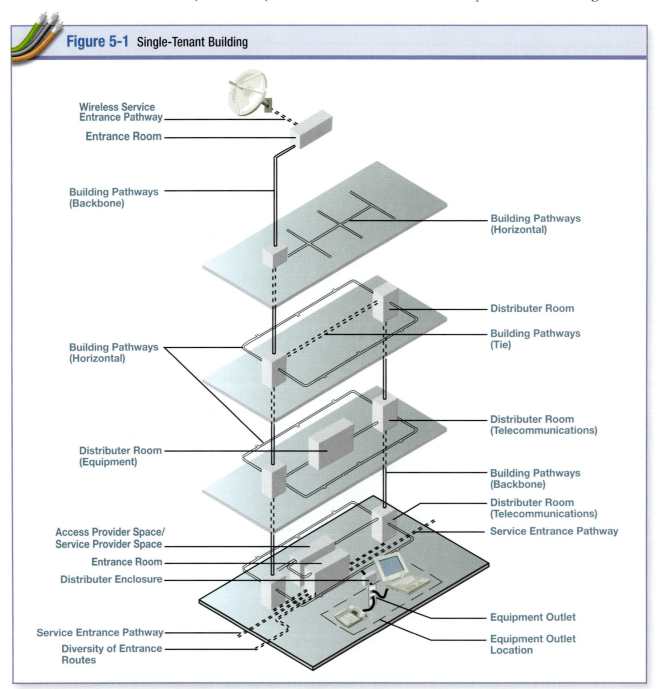

Figure 5-1 Single-Tenant Building

Wireless Service Entrance Pathway

Entrance Room

Building Pathways (Backbone)

Building Pathways (Horizontal)

Building Pathways (Horizontal)

Distributer Room

Building Pathways (Tie)

Distributer Room (Telecommunications)

Distributer Room (Equipment)

Building Pathways (Backbone)

Distributer Room (Telecommunications)

Access Provider Space/ Service Provider Space

Entrance Room

Distributer Enclosure

Service Entrance Pathway

Equipment Outlet

Service Entrance Pathway

Diversity of Entrance Routes

Equipment Outlet Location

Figure 5-1. *There are several different pathways and spaces recognized in the ANSI/TIA-569-E standard for a single-tenant building.*

Also, the changing nature of technology and the ease with which the tenant can adapt to these changes may be of critical concern to the tenant. However, whether the building is owner-occupied or tenant-occupied, the design of a building's communications infrastructure should allow for the greatest flexibility and for future growth and changes in technology. **See Figure 5-2.** Having a standard set of guidelines allows architects, engineers, and specifiers to develop designs with the greatest flexibility.

SEPARATION BETWEEN CABLE AND POWER

While the original ANSI/TIA-569 Standard listed specific minimum separation distances between telecommunications cabling pathways and electrical power

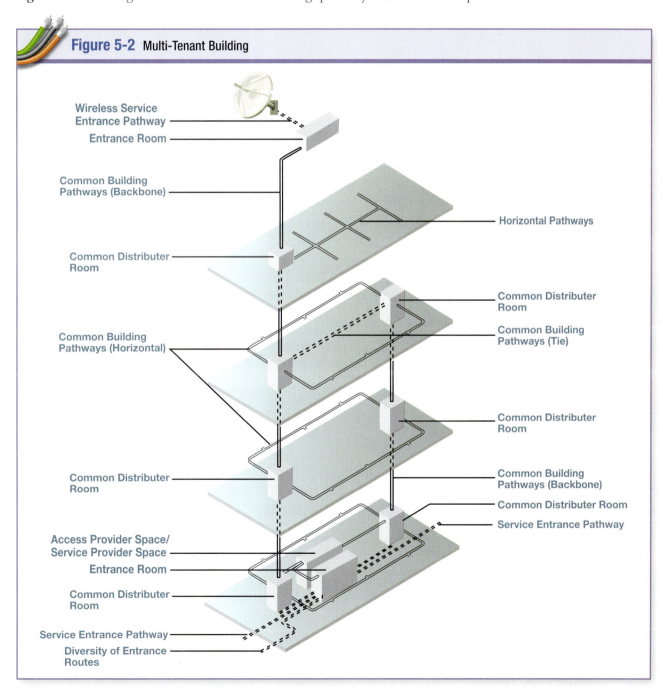

Figure 5-2 Multi-Tenant Building

Wireless Service Entrance Pathway

Entrance Room

Common Building Pathways (Backbone)

Horizontal Pathways

Common Distributer Room

Common Distributer Room

Common Building Pathways (Tie)

Common Building Pathways (Horizontal)

Common Distributer Room

Common Distributer Room

Common Building Pathways (Backbone)

Common Distributer Room

Service Entrance Pathway

Access Provider Space/ Service Provider Space

Entrance Room

Common Distributer Room

Service Entrance Pathway

Diversity of Entrance Routes

Figure 5-2. There are many different pathways and spaces recognized in the ANSI/TIA-569-E Standard for a multi-tenant building.

pathways, the current standard does not. It recommends that co-installation of telecommunications cable and power cable be governed by the applicable electrical code for safety. For minimum separation requirements of electrically conductive telecommunications cable from typical branch circuits (120/240 volts, 20 amperes), 805.133(A)(1) and (A)(2) of the *National Electrical Code (NFPA 70)* shall be applied. Examples of this include:

1. Separation from power conductors
2. Separation and barriers within raceways
3. Separation within outlet boxes or compartments

The ANSI/TIA-569-E Standard recommends that the following precautions be observed in order to minimize noise coupling between power systems and telecommunications wiring:

- Increase physical separation between power and telecommunications wiring.
- Maintain electrical branch-circuit line, neutral, and grounding conductors close together (for example, twisted, sheathed, taped, or bundled together) to minimize inductive coupling into telecommunications cabling.
- Use surge protectors in electrical branch circuits to further limit the propagation of electrical surges.

- Use fully enclosed grounded metallic raceway or grounded conduit or install telecommunications cables close to a grounded metallic surface to further limit inductive noise coupling.

HORIZONTAL PATHWAYS

Horizontal pathways consist of the supporting structures for concealing and/or supporting the cables placed between telecommunications rooms and work areas. There are several types of horizontal pathways.

Ceiling Methods

- Home run
- Zone
- Raceway
- Poke through
- Cable tray and cable runways

Floor Methods

- Underfloor duct system
- Cellular floor
- Raised floor (access floor)
- Underfloor conduit

Ceiling Methods

Ceiling distribution systems shall provide full access to all horizontal pathways. Fixed-style ceilings may be used if one of the following conditions exists:

- A crawl or walk space exists that provides full access to the distribution system.

Tech Fact

In the past, the following methods were allowed when installing cables without a means of continuous support (that is, using part of a suspended ceiling system for support).

- Appropriate cable supports were allowed to be attached to suspended ceiling support rods and wires.
- Appropriate cable supports were allowed to be mounted on the T-bar rail of a suspended ceiling.

These methods are no longer allowed. The current acceptable method of non-continuous support is found in TIA-569-E-9.8:

Non-continuous supports shall be located at intervals not to exceed 1.5 m (5 ft). Non-continuous supports shall be selected to accommodate the immediate and anticipated quantity, weight, and performance requirements of cables.

Steel, masonry, independent rods, independent support wires or other structural parts of the building shall be used for cable support attachment points up to the total weight for which the fastener is approved. Rods or wires that are currently employed for other functions (for example, suspended ceiling grid support) shall not be utilized as attachment points for non-continuous supports.

Note: A weight of one kilogram (2.2 lb) (or 0.7 kg/m) with spacing of support wire/rod at 1.5 meters (5 ft) is equivalent to a bundle of sixteen 4-pair 24 AWG UTP cables, including fasteners.

Figure 5-3 Home-Run Method

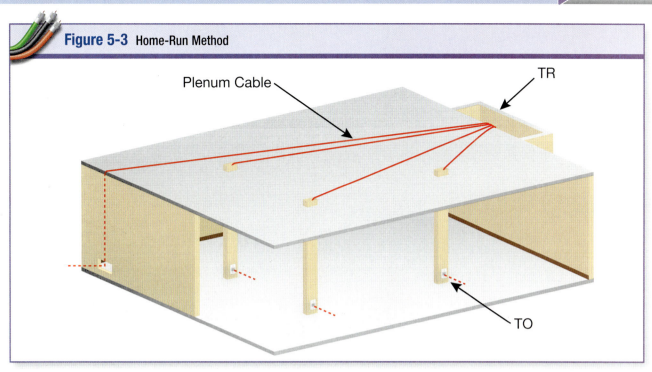

Plenum Cable

TR

TO

Figure 5-3. *In the home-run method, cables are routed directly from the telecommunications room to the outlets.*

- If locking tiles exist, they are modified to allow easy removal and replacement.
- A closed pathway exists or is provided.

The cable supports used with open ceiling systems must be a minimum of three inches above the ceiling grid. The installed materials in open ceilings must be properly rated (plenum or non-plenum), and comply with all applicable building and electrical codes required in an environmental air-handling space. All cable supports shall be structurally independent of the suspended ceiling support system and framework.

Ceiling Home-Run Method
The home-run method may involve using plenum cables that are routed directly from the serving telecommunications room to the telecommunications outlets at the work areas. **See Figure 5-3.** This method is economical and offers the most flexibility for distributing cables in a ceiling. It also minimizes crosstalk coupling among cables, since cables are not in the same pathway and their spacing is varied. In this method of home-run cabling, the cables must be routed and secured in a neat and workman-like manner.

The TIA-569-E Standard requires that cables run in open ceilings be supported at least every five feet by adequate open-top cable supports (J-hooks).

The following should be considered when installing cable above ceilings.

1. Inaccessible ceiling areas, such as lock-in type ceiling tiles, drywall, or plaster, shall not be used as distribution pathways.
2. Ceiling tiles shall be of the removable or lay-in type.
3. Adequate and suitable space shall be available in the ceiling area for the contemplated layout.

The design of the pathway above the ceiling shall provide a suitable means and method for supporting cables. The cables shall not be laid directly on the ceiling tile or ceiling rails, and a minimum of 75 millimeters (3 in) of clear vertical space shall be available above the ceiling tiles for the horizontal cabling and pathway.

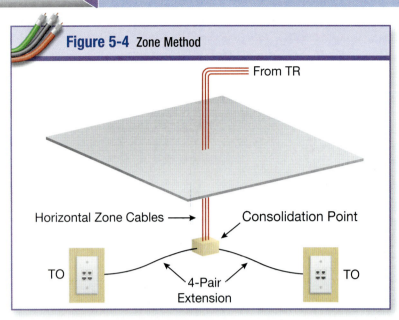

Figure 5-4 Zone Method

From TR

Horizontal Zone Cables ⟶   Consolidation Point

TO   TO

4-Pair Extension

Figure 5-4. The zone method involves zone cables and flexible extension cables routed to a consolidation point.

Ceiling Zone Method

Zone cabling is a useful option for supporting open office work areas. **See Figure 5-4.** Open office design is a newer practice based on modular furniture, flexible partitions, and project workgroups. The ceiling zone method divides horizontal cabling into two sections:

1. A permanent section from the TR to a consolidation point (the zone cables)
2. An adjustable or flexible section from the consolidation point to the TOs (extension cables)

The consolidation point establishes a cluster or group of work areas (zone). Consolidation points must house and otherwise protect the interconnection of the zone and extension cables.

Ceiling Raceway Method

TIA-569-E defines *raceway* as any enclosed channel designed for holding wires or cables. A ceiling raceway system utilizes a raceway that may consist of open or closed metal trays suspended in the ceiling area from the floor above. The *NEC* refers to this type of raceway as a *metal wireway* (see Article 376). While this type of system is seldom used, it may be installed in larger buildings or where the distribution system is

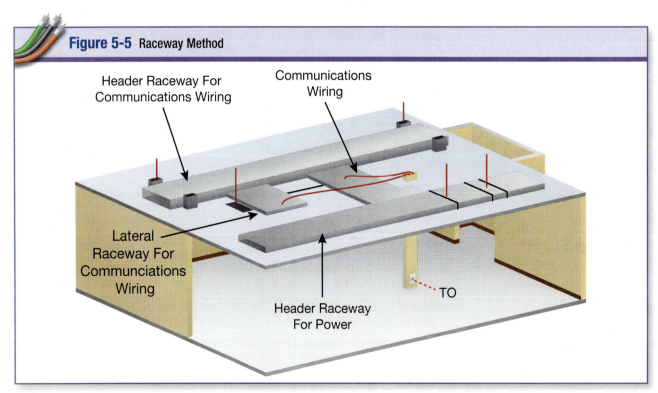

Figure 5-5 Raceway Method

Header Raceway For Communications Wiring

Communications Wiring

Lateral Raceway For Communciations Wiring

Header Raceway For Power

TO

Figure 5-5. Raceway methods are often used in larger buildings to allow for a wider distribution.

complex enough to demand the extra support. In this design, a header raceway is used to bring cables into the desired area. Lateral raceways branch off from the header raceway to provide services to the floor space below. **See Figure 5-5.** Cables are then run to utility columns or partition walls through short lengths of flexible conduit and terminated in TOs at the work areas.

Conduit may be used as an approved above-ceiling horizontal raceway method. However, because of its lack of flexibility once it is installed, conduit should be considered only when it is required by code, outlet locations are permanent, device densities are low, special mechanical protection is required, or flexibility is not required. Flexible metal conduit is not recommended, but when used, the length should be limited to six meters (20 feet) for each run, and the conduit selected should minimize cable abrasion.

Innerduct (also known as subduct) is typically a nonmetallic pathway within a pathway and may be used in accordance with appropriate codes for installation of cable to facilitate subsequent placement of additional cable in a single pathway.

When installing conduit as a horizontal raceway, a continuous length of conduit shall not exceed 100 feet between pull points and no section of conduit shall contain more than two 90° bends, or the equivalent. Conduit sizing information is referenced in the *National Electrical Code (NEC)*.

Poke-Through Method

The poke-through method requires an approved poke-through device. This device is placed in a hole in the floor, and cables are "poked" up from the ceiling space below. **See Figure 5-6.** Approved firestopping methods must be implemented around the cables where they pass through the floor.

TIA-569-E defines a *poke-through device* as an assembly that allows through-penetration of floor decking with telecommunications cables, power cables, or both, while maintaining the fire-rating

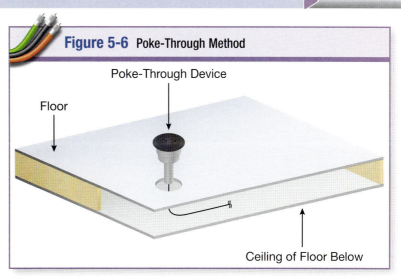

Figure 5-6 Poke-Through Method

Figure 5-6. The poke-through method requires an approved poke-through device.

integrity of the floor. A *poke-through system* is defined as a poke-through device installed in a penetration through a fire-resistant floor structure.

If the space below the floor is considered a plenum space (a ceiling plenum for the floor below), the poke-through device shall be suitable for use in air-handling spaces in accordance with 300.22(C) of the *NEC*.

Per TIA-569-E, before proceeding with the installation of a poke-through device or system, the fire rating of the floor will need to be determined. The intended poke-through device shall be listed for that purpose. The requirements of the appropriate fire resistance directory (published by nationally recognized testing laboratories) for allowable density of poke-through devices and minimum spacing between devices shall be followed. A licensed structural engineer shall approve the location and density of the poke-through devices, and all locations and sizes of installed poke-through devices shall be documented in building records. The manufacturer shall provide information about the allowable cross-sectional area of cables for each device, and all abandoned poke-through holes shall be properly firestopped.

Floor Methods

The basic design characteristics of floor distribution systems require that floor panels, if used exclusive of their covering, shall be in compliance with all minimum flame spread ratings per all applicable codes and regulations. Defined, dedicated pathways shall be used in all flooring system applications, and when a flooring system is used for handling environmental air, it shall be in compliance with all applicable building and electrical codes.

There are numerous methods that may be used for floor pathways. How-ever, there are advantages and disadvantages to each system. **See Figure 5-7.**

Underfloor Duct Method

An *underfloor duct system* consists of a series of metal distribution channels, often enclosed in concrete with metallic feeder troughs. Depending on communications and power cabling requirements, the thickness of the floor, and the floor space to be occupied, underfloor duct systems may be one or two levels. A one-level system consists of distribution and feeder ducting on a single plane. **See Figure 5-8.** A two-level system consists of a double layer with the feeder duct

Figure 5-7 Floor Pathway Summary

	Floor Pathways	
Methods	**Advantages**	**Disadvantages**
Underfloor Duct	• Provides mechanical protection • Reduces electrical interference • Increases security • Preserves appearance • Reduces safety hazards	• Expensive to install; installation is necessary before completion of building construction • Requires special treatment of openings for service fittings in carpeted areas • Adds floor weight • Limits flexibility
Cellular Floor	• Provides mechanical protection • Reduces electrical interference • Increases security • Preserves appearance • Reduces safety hazards • Somewhat more cable capacity	• Expensive to install • Installation is necessary before underfloor duct completion of construction • Limits flexibility
Raised Floor	• Flexible • Easy to install • Ample cable capacity • Easy access to wiring • Easy to fireproof	• Expensive to install • Adds floor weight • "Sounding-board" effect • Poor control over cable runs • Decreases telecommunications room height if it is not planned effectively • Requires plenum cable when space under floor is used as an air plenum
Underfloor Conduit	• Preserves appearance • Low initial installation cost	• Limits flexibility

Figure 5-7. Each floor pathway has advantages and disadvantages.

running underneath the distribution level.

Cellular Floor Method

Cellular floors consist of a series of channels through which cables pass. **See Figure 5-9.** Floor cells constructed of steel or concrete provide a ready-made raceway for distribution of power and telecommunications cables. Alternating power and communications cells can provide a flexible layout.

Depending on the floor structure, distribution cells are made of either steel or concrete. In both cases, header ducts are used as feeders to pull cables from distribution cells to telecommunications rooms.

Raised Floor Method

The *raised floor* (also called *access floor*) consists of square plates that rest on locking aluminum or steel pedestals attached to the building floor. **See Figure 5-10.** The plates typically consist of a steel-bottom plate adhesively bonded to a laminated wood core covered by cork, carpet tiles, or vinyl tiles. Any square is removable for access to the cables beneath.

When the space beneath the floor is used as an air plenum, the installation of plenum-rated cables may be necessary, depending upon local code requirements. The TIA-569-E Standard recommends that the minimum finished floor height should be 300 millimeters (12 in) but never less than 150 millimeters (6 in) in an equipment room. In general office spaces, a minimum of 200 millimeters (8 in) is recommended.

Underfloor Conduit Method

An *underfloor conduit system* is made up of many metal pipes radiating from a serving room to potential workstation locations in the floors, walls, or columns of an office space. **See Figure 5-11.** If enough TOs are installed, the system is suitable for buildings with relatively stable terminal locations, such as department stores, banks, and small medical buildings.

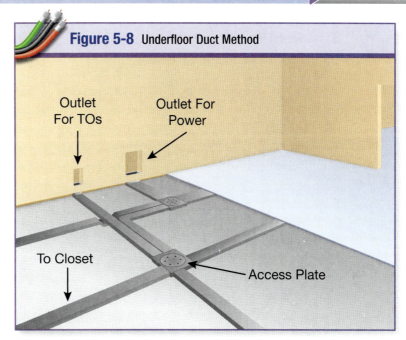

Figure 5-8. *The underfloor duct method involves the use of a system of metal distribution channels beneath the floor of a building.*

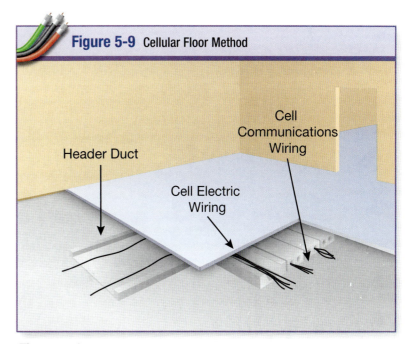

Figure 5-9. *Cellular floors have a built-in system of channels through which cable can be distributed.*

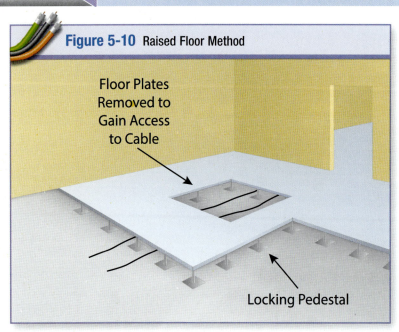

Figure 5-10 Raised Floor Method

Floor Plates Removed to Gain Access to Cable

Locking Pedestal

Figure 5-10. Cable installed using the raised-floor method can be accessed by removing any of the square plates that make up the floor.

FURNITURE PATHWAYS

The TIA-569-E Standard requires that:

- Furniture used for telecommunications cabling shall provide a minimum (straight) pathway cross-sectional area of 1.5 square inches.

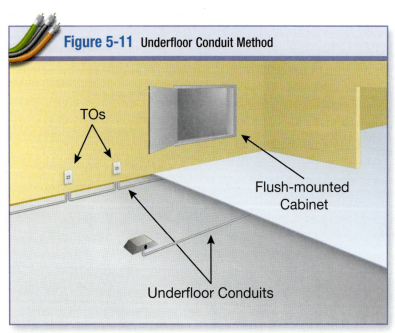

Figure 5-11 Underfloor Conduit Method

TOs

Flush-mounted Cabinet

Underfloor Conduits

Figure 5-11. The underfloor conduit method uses a system of conduit running throughout the building to distribute cable.

- Fish-and-pull installation should not be used except when required by furniture pathway characteristics.
- The minimum size pathway shall not force the cable into a bend radius less than one inch under condition of maximum fill.
- Sufficient space shall be provided so that bend radius requirements are not violated in termination spaces.

It is additionally suggested that telecommunications outlets/connectors be mounted in such a way that they do not reduce the required pathway cabling capacity.

INTRA-BUILDING BACKBONE PATHWAYS

The backbone pathway consists of the portion of the pathway system that permits the placing of backbone cables between the entrance location and all cross-connect points within a building and between buildings. There are four commonly used pathway types for intra-building backbone cables:

- Sleeve
- Slot
- Cable tray
- Conduit

Sleeve Method

The use of sleeves is the most common method for routing intra-building backbone cables among vertically stacked telecommunications rooms. Sleeves are usually four inches in diameter and are made of PVC or steel. They are placed in concrete floors as they are being poured. **See Figure 5-12.**

The TIA-569-E Standard specifies that a minimum of five 4-inch sleeves (or 4-inch trade size conduits) should be installed to serve up to 4,000 square meters (40,000 ft^2), and there should be one additional sleeve per every additional 4,000 square meters (40,000 ft^2) of usable floor space. Location and configuration of sleeves shall be approved by a licensed structural engineer. Slots may be used in place of sleeves where

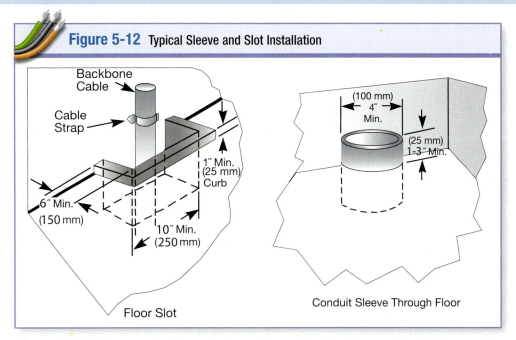

Figure 5-12 Typical Sleeve and Slot Installation

Figure 5-12. Sleeves and slots are both used for routing backbone cables between vertically stacked telecommunications rooms.

large numbers of cables are expected. **See Figure 5-13.**

When telecommunications rooms are aligned vertically within a multi-story building, thought should be given to installing a means of cable pulling above and inline with the sleeves or slots at the upper-most room of each vertical stack. A steel anchor pulling iron or eye embedded in the concrete is an example.

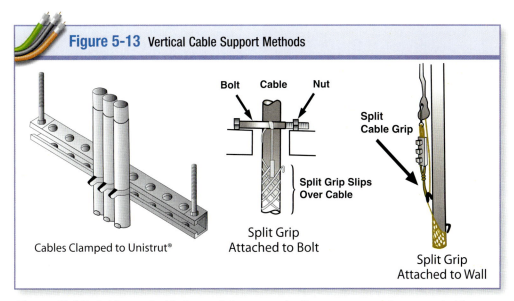

Figure 5-13 Vertical Cable Support Methods

Figure 5-13. Cables routed between telecommunications rooms can be supported in a variety of ways.

Slot Method

Slots are an alternative to sleeves. A *slot* is an opening (usually rectangular) in a wall, floor, or ceiling that allows the passage of cables and wires. **See Figure 5-14.** As with the sleeve method, cables are tied off to support hardware on each floor. Vertical racks may be placed on the wall adjacent to the slots in order to support large quantities of cables.

Slots are very flexible, allowing any combination of cable types and sizes. However, they are more expensive than sleeves and they are more difficult to firestop.

Slots are typically located flush against the wall within a space, and should be designed at a depth (the dimension perpendicular to the wall) of 150 to 600 millimeters (6–24 in), giving preference to narrower depths wherever possible. The location and configuration of the slot(s) shall be approved by a structural engineer.

The size of the pathway using slots should be one slot sized at 0.04 square

meters (60 in²) for up to 4,000 square meters (40,000 ft²) of usable floor space served by that backbone system. As with the sleeve method, the slot should be increased in size by 0.04 square meters (60 in²) with each 4,000-square-meter (40,000 ft²) increase in useable floor space.

Cable Tray Method

Cable tray and *cable runway* are cable support systems that may be used for above-ceiling pathway applications or in-access floor pathway applications. **See Figure 5-15.** They are support mechanisms used to route and support telecommunications cable or power cable. They are typically equipped with sides that allow cables to be placed within the sides over their entire length.

TIA-569-E indicates that cable trays and cable runways are structures with pre-fabricated components for supporting and routing cables or conductors that are pulled or laid in place after the pathway has been installed as a complete system.

The *NEC* defines a cable tray system as "a unit or assembly of units or sections and associated fittings forming a structural system used to securely fasten or support cables and raceways."

Cable runway (ladder rack) supports backbone and horizontal cables between the point of entry/exit into the telecommunications or equipment room and the cross-connects on the rack or cabinet. **See Figure 5-16.** These integrated cable pathway systems route cable within and outside of the equipment room, under raised floors, and above the ceiling.

Cable tray and runway come in the following configurations:

1. Ladder
2. Solid-bottom
3. Trough
4. Channel
5. Wire mesh
6. Single rail

The method of installation and support for cable tray and cable runway

Figure 5-14 Slot Method

Figure 5-14. *The slot method involves an opening in the floor or ceiling through which cables are routed.*

systems should be determined in accordance with the manufacturer's maximum recommended load capacity for a given span.

There are three basic methods for supporting cable tray and/or cable runways. One method is to use cantilever brackets mounted on a wall. Another is to use a trapeze or individual rod supports mounted to the ceiling structure. The third method is to mount the cable tray or cable runway directly to the floor. Cable tray supports should be located within 600 millimeters (24 in) on each side of any connection to a bend, tee, or cross.

See the NEMA-VE2 *Metal Cable Tray Installation Guidelines* for more information regarding additional cable tray support and installation.

Power and telecommunications cables may share the same cable tray provided that the tray is divided with a barrier to allow for physical separation between power cable and conductive telecommunications cables. Power and telecommunications cables installed in tray shall be installed per the *NEC*.

TIA-569-E states that a minimum of 200 millimeters (8 in) of access headroom shall be provided and maintained above a cable tray system or cable runway; the recommended access headroom is 300 millimeters (12 in). During the planning and installation of the tray system, care shall be taken to ensure that other building components (for example, sprinkler pipes or air conditioning ducts) do not restrict access to the tray.

When running cable trays and cable runways within the ceiling and into the telecommunications room, the tray or runways shall penetrate into the room 25 to 75 millimeters (1–3 in) without a bend and must be 2.4 meters (8 ft) above the level of the finished floor. The TIA-569-E pathway entry requirements for cable runway into the telecommunications room aim to prevent partial bend transitions through the wall and ensure that the cable is at a height that may be fed to termination fields without

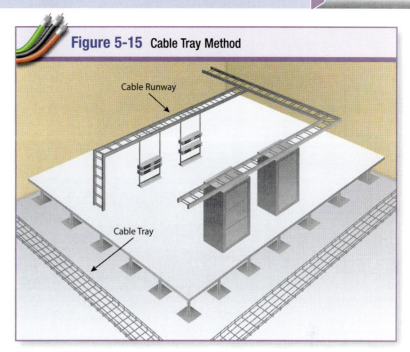

Figure 5-15. *Cable trays can be used to support telecommunications cable.*

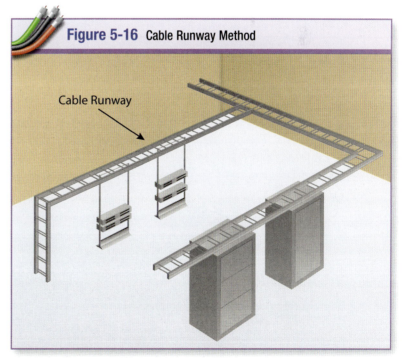

Figure 5-16. *Cable runways (ladder rack) can be attached to either the wall or the ceiling.*

Figure 5-17 Ceiling Pathway Summary

| | Ceiling Pathway | |
Methods	Advantages	Disadvantages
Home-Run	• Most flexible • Economical	• May be more expensive initially than zone, depending on number of cables required
Zone	• Somewhat flexible • Economical	• Limits flexibility when conduit is used, depending on conduit size
Raceway	• Provides mechanical protection and support • Effective for large installations • Offers the greatest protection	• Expensive to install and may limit flexibility • Can add excessive weight in ceiling for the cables
Poke-Through	• A versatile system for adding or up-grading telecommunications and AV capabilities in an existing multi-story concrete slab building • Allows for design versatility in open office space	• Interrupts work activity • Requires core drilling of the existing floor structure • Cannot be used for slab-on-grade applications
Cable Tray	• Lower cost to install than raceway systems • Reduced cable pulling costs	• Additional bonding and grounding requirements • Fiber optic cable installations may require special attention to avoid crush loads from copper cabling and micro bending caused by ladder rungs • Leaves cables exposed • Is difficult to firestop • May not be acceptable in terms of its effect on a building's interior appearance

Figure 5-17. Each ceiling pathway has advantages and disadvantages.

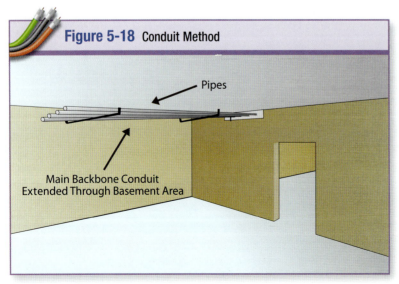

Figure 5-18 Conduit Method

Pipes

Main Backbone Conduit Extended Through Basement Area

Figure 5-18. Conduits create paths to allow cable to be distributed through a building.

interfering with equipment racks or back panels. See **Figure 5-17**.

Recall that cable trays are metal assemblies that resemble ladders. They may be attached to walls for vertical runs and to ceilings for horizontal runs. Cables are laid along the rack and occasionally tied to its horizontal support members. Cable trays allow easy placing and eliminate the problems presented by pulling through conduits.

The tray method leaves cables exposed, is difficult to firestop, and may not be acceptable in terms of its effect on a building's interior appearance.

Conduit Method

Conduits are an acceptable pathway for intra-building backbone cables. **See Figure 5-18**. However, they may not be the preferred method because of cost and lack of flexibility after installation. Conduits allow cables to be pulled in paths with horizontal offsets between rooms on adjacent floors.

Conduits provide mechanical protection for cables routed in shafts, through ceilings, and in basements.

The TIA-569-E Standard contains tables for conduit fill, pull box sizing, and splice box sizing. This standard requires that:

- No section of conduit shall be longer than 100 feet or contain more than two 90° bends between pull points or boxes.

Example

Determine if three 300-pair, 1.3-inch diameter cables may be placed in a 4-inch conduit.

1. Determine the area of the cable (diameter to area conversion); 1.30 inch diameter is 1.34 square inches (8.71 cm²). **See Figure 5-19**.
2. Multiply the area of one cable (1.34) by three. 3 × 1.34 = 4.02 square inches
3. Determine the area of four-inch conduit. Internal diameter is 4.03 and area is 12.83 square inches (82.68 cm²). **See Figure 5-20**.
4. Refer to the column with three cables or 40% fill, which shows the maximum fill to be 5.13 square inches (33.07 cm²).

Since the area for three 300-pair, 1.3-inch diameter cables (4.02 in²) is less than the 5.13 inches squared allowed by the table look-up, the three cables may be placed in a 4-inch conduit. **See Figure 5-21**.

When backbone cables are placed in conduit runs requiring bends, a long bend radius must be used. **See Figure 5-22**.

Figure 5-19 Diameter to Area Conversion

Diameter to Area Conversion

Diameter	Area (sq. in.) (0.79D²)	(cm²)	Diameter	Area (sq. in.) (0.79D²)	(cm²)	Diameter	Area (sq. in.) (0.79D²)	(cm²)
0.3	0.07	.455	2.2	3.82	24.83	4.1	13.28	86.32
0.4	0.13	.845	2.3	4.18	27.17	4.2	13.93	90.54
0.5	0.20	1.30	2.4	4.55	29.57	4.3	14.61	94.96
0.6	0.28	1.82	2.5	4.94	32.11	4.4	15.29	99.38
0.7	0.39	2.53	2.6	5.34	34.71	4.5	16.00	104.00
0.8	0.51	3.31	2.7	5.76	37.44	4.6	16.72	108.68
0.9	0.64	4.16	2.8	6.19	40.23	4.7	17.45	113.42
1.0	0.79	5.14	2.9	6.64	43.16	4.8	18.20	118.30
1.1	0.96	6.24	3.0	7.11	46.21	4.9	18.97	123.30
1.2	1.14	7.41	3.1	7.59	49.33	5.0	19.75	128.37
1.3	1.34	8.71	3.2	8.09	52.58	5.1	20.55	133.57
1.4	1.55	10.75	3.3	8.60	55.90	5.2	21.36	138.84
1.5	1.78	11.78	3.4	9.13	59.34	5.3	22.19	144.23
1.6	2.02	13.13	3.5	9.68	62.92	5.4	23.04	149.76
1.7	2.28	14.82	3.6	10.24	66.56	5.5	23.90	155.35
1.8	2.56	16.64	3.7	10.82	70.33	5.6	24.77	161.00
1.9	2.85	18.52	3.8	11.41	74.16	5.7	25.67	166.85
2.0	3.16	20.54	3.9	12.02	78.13	5.8	26.58	172.77
2.1	3.48	22.62	4.0	12.64	82.16	5.9	27.50	178.75
						6.0	28.44	184.86

Figure 5-19. *Conversion charts are used to calculate conduit fill.*

- Minimum bend radius shall be six times the internal diameter for two-inch and smaller conduits, and 10 times the inside diameter for conduits larger than two inches in diameter.
- All conduit ends are to be reamed and bushed.
- Conduits that penetrate through floors are to be terminated one to three inches above the floor.

PULL AND SPLICE BOX RECOMMENDATION

Pull boxes shall be used for the following purposes:

- Fishing the conduit run
- Pulling the cable to the box and looping the cable to be pulled into the next length of conduit; this is usually done only with smaller cables and not with cables with a

Figure 5-20 EMT Conduit Fill for Backbone Cable

| | | | Maximum Recommended Occupancy | | | Minimum Bend Radius | |
| | | | A | B | C | D | E |
Size Diameter	Internal Designator	Total Area 100%	1 Cable 53% Fill	2 Cables 31% Fill	3 Cables and over	Steel Layers within 40% Fill	Other Sheaths
mm (in)	mm (in)	mm (in²)	mm² (in²)	mm² (in²)	mm² (in²)	mm (in)	mm (in)
21 (¾)	20.9 (.82)	343 (.53)	182 (.28)	106 (.17)	137 (.21)	210 (8)	130 (5)
27 (1)	26.6 (1.05)	556 (.86)	295 (.46)	172 (.27)	222 (.35)	270 (10)	160 (6)
35 (1 ¼)	35.1 (1.38)	968 (1.50)	513 (.79)	300 (.46)	387 (.60)	350 (13)	210 (8)
41 (1 ½)	40.9 (1.61)	1314 (2.04)	696 (1.08)	407 (.63)	526 (.81)	410 (15)	250 (9)
53 (2)	52.5 (2.07)	2165 (3.36)	1147 (1.78)	671 (1.04)	866 (1.34)	530 (20)	320 (12)
63 (2 ½)	69.4 (2.73)	3783 (5.86)	2005 (3.11)	1173 (1.82)	1513 (2.34)	630 (25)	630 (25)
78 (3)	85.2 (3.36)	5701 (8.85)	3022 (4.69)	1767 (2.74)	2280 (3.54)	780 (30)	780 (30)
91 (3 ½)	97.4 (3.83)	7451 (11.55)	3949 (6.12)	2310 (3.58)	2980 (4.62)	900 (35)	900 (35)
103 (4)	110.1 (4.33)	9521 (14.75)	5046 (7.82)	2951 (4.57)	3808 (5.90)	1020 (40)	1020 (40)

NOTES
1. Column A is used when one cable is to be placed in a conduit.
2. Column B is used when two cables share a conduit. The percentage fill is applied to straight runs with nominal offset equivalent to no more than two 90° bends.
3. Column C is used when three or more cables share a conduit.
4. Column D indicates a bend of 10 times the conduit diameter for cable sheaths equipped with steel tape in the sheath.
5. Column E indicates a bend of six times the conduit diameter up to and including 53 (2) trade size, and 10 times the diameter above 53 (2) trade size conduit.
6. The number of cables that can be installed in a conduit can be limited by the allowed maximum pulling tension of the cables.
7. For large diameter cables, conduit fill is a factor of cable pulling tension.

Figure 5-20. Conduit fill charts aid in determining the number of cables that can be placed in a conduit.

diameter of 64 millimeters (2.5 in) or greater

Pull boxes shall not be used for splicing cable. Splice boxes are intended for splicing in addition to pulling cable.

Pull or splice boxes shall be placed in an exposed manner and location, and they shall be readily accessible. Pull or splice boxes shall not be placed in a fixed false ceiling space unless immediately above a suitably marked hinged panel.

A pull or splice box shall be placed in a conduit run where:

- The length is over 30 meters (100 ft)
- There are more than two 90° bends or the equivalent
- There is a reverse (U-shaped) bend in the run.

Boxes shall be placed in a straight section of conduit and not used in lieu of a bend. The corresponding conduit ends should be aligned with each other. Conduit fittings shall not be used in place of pull boxes. **See Figure 5-23.**

If slip sleeves, gutters, or open sections of conduit are used instead of pull boxes, make the opening as long as the pull box specified.

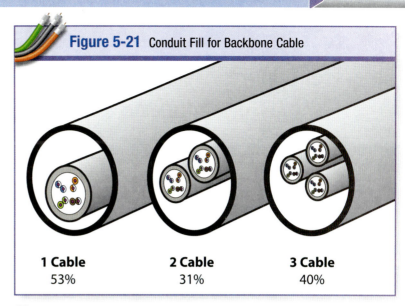

Figure 5-21 Conduit Fill for Backbone Cable

1 Cable	2 Cable	3 Cable
53%	31%	40%

Figure 5-21. Backbone cable conduit fill percentages are calculated based on the number of cables to be placed in each conduit.

Figure 5-22 Minimum Bend Radius

Size of Conduit (inches)	Cross Section (sq. inch)	Minimum Radius of 90° Conduit Bend	
		Inches	mm
0.75	0.53	5	130
1.00	0.87	6	160
1.25	1.51	8	210
1.50	2.05	10	250
2.00	3.39	12	320
2.50	4.82	25	630
3.00	7.45	31	780
3.50	9.96	36	900
4.00	12.83	40	1,020
5.00	20.15	50	1,280
6.00	29.11	60	1,540

Figure 5-22. Charts can be used to aid in determining the minimum bend radius.

Figure 5-23 Minimum Pull and Splice Box Sizes

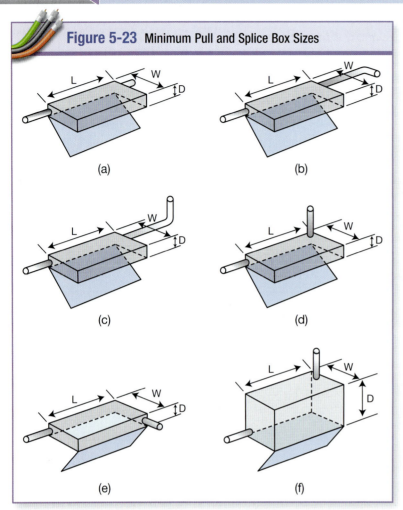

(a)

(b)

(c)

(d)

(e)

(f)

Figure 5-23. For pulling cable through pull and splice boxes, figures (a), (b), and (c) are acceptable; (d), (e), and (f) are not. If a 90° bend is required at the pull box, the configurations shown in (b) and (c) are preferred.

When determining the size of a pull box that is needed for a specific installation, a chart must be used. **See Figure 5-24.** Assume a pull box is needed and there are two 1¼-inch conduits run to the pull box (one in and one out). The minimum pull box size required is six inches wide by 20 inches long by three inches deep. If any additional conduits are added to the pull box, additional width is required for the pull box. For example, if there are two additional 1¼-inch conduits added to the pull box with a starting size of 6 inches × 20 inches × 3 inches, based on the chart, an additional three inches

must be added to the width of the pull box for each additional conduit. The minimum size pull box required is now 12 inches wide by 20 inches long by 3 inches deep.

FIRESTOPPING

Firestopping is a system used to prevent the spread of fire, smoke, toxic fumes, or water from passing through fire-rated walls, floors, and floor/ceiling assemblies by reestablishing the fire-resistance rating of the barrier.

In order to contain fire, smoke, and toxic fumes, and to prevent them from spreading throughout a building, fire-rated doors and walls are required in commercial buildings. Building codes require these fire-rated doors, floors, and walls to isolate areas where hazards are likely to exist. However, it is often necessary to penetrate these fire barriers in order to get electrical, communications, and mechanical services into areas where they are needed. Firestopping products are designed to restore the integrity of fire-rated walls and floors after they have been penetrated.

During construction, it is the responsibility of the general contractor to see that all penetrations are properly firestopped. However, after occupancy, the need to penetrate fire-rated floors and walls continues. Additional wiring creates new holes in existing fire barriers which then must be firestopped. Types of openings that to be need firestopped include:

- Through penetrations
- Membrane penetrations
- Fire resistive construction joints
- Perimeter joints

A *through-penetration firestop system* consists of the fire-rated wall or floor, the penetrating item (cable, conduit, tray, etc.), and the firestop material. A *membrane penetration* is an opening made through one side of a fire-rated wall, floor, or floor/ceiling assembly. An example of this is a cable penetrating one side of a fire wall to supply a work area outlet.

Before a firestopping system can be installed, there are some things that must be done. First, the hourly rating of the barrier must be determined. Next is the

Figure 5-24 Pull and Splice Box Sizing

Nominal Conduit Size (inches)	Pull Box Sizes for Two Conduits (inches)			Increase Width For Each Additional Conduit (inches)
	Configurations (a) or (b) or (c)			
	W (inches)	L (inches)	D (inches)	
3/4	4	12	3	2
1	4	16	3	2
1 - 1/4	6	20	3	3
1 - 1/2	8	27	4	4
2	8	36	4	5
2 - 1/2	10	42	5	6
3	12	48	5	6
3 - 1/2	12	54	6	6
4	15	60	8	8
	Splice Box Sizes for Two Conduits (inches)			
	Configurations (a) or (b) or (c)			
	W (inches)	L (inches)	D (inches)	
3/4	10	30	3	3
1	12	32	4	3
1 - 1/4	14	36	5	4
1 - 1/2	18	39	6	4
2	20	42	7	5
2 - 1/2	24	48	8	6
3	30	54	9	6
3 - 1/2	36	60	10	7
4	42	66	11	7

Figure 5-24. Splice box and pull box size is determined by the number of conduits entering the box as well as the size of the conduits.

proper selection of the listed method/product to seal the penetration that will match or exceed the fire rating of the barrier. Once the fire rating and listed method have been determined, there must be approval from the authority having jurisdiction (inspector) before work begins. For rated systems, the complete firestop system is tested and listed; it is not just the firestop material alone that must be checked. TIA-569-E Annex A.6 contains a design consideration checklist that may be used in meeting the requirements of a properly installed firestopping system.

A variety of firestopping methods and materials are used to prevent the spread of fire and smoke for the systems listed above. Typical firestopping materials are:

- Mechanical systems
- Non-mechanical systems
  - Putty
  - Caulks
  - Cementitious mortars and grouts
  - Intumescent wraps/sheets
  - Silicone foams
  - Pillows, bags, blocks

Mechanical Firestop Systems

Mechanical firestops consist of metal frames containing elastomeric modules fitted to the cables with some method of applying compression to the assembly. When properly installed, mechanical systems are outstanding for their ability to withstand shock and vibration and their ability to provide airtight, watertight, chemical-resistant seals. In addition, they are easily adaptable when cabling changes have to be made. However, mechanical systems require design layouts and must be carefully planned.

Mechanical firestopping systems consist of components that are shaped to fit around cables and conduits. These components are fitted around penetrating cables and arranged to fit within a frame. See Figure 5-25.

Non-Mechanical Firestop Systems

Non-mechanical firestop systems are available in a variety of forms to fit many different situations. They were developed to fit irregular shaped openings and surround off-centered items.

Putty is a popular firestop wherever frequent cable revisions are likely. Most putties are intumescent, meaning that they expand or swell when exposed to heat. They are available in bars, sticks, pads, or tubes. Putty has no hazardous or noxious fumes, and provides an immediate firestop; it is ideal for retrofit work in occupied areas. No special storage precautions are necessary, and putty can be installed over a wide temperature range. Putty installations are also easily inspected.

Putty shall be installed in conjunction with ceramic fiber, rock wool filler, or other approved material. **See Figure 5-26.** Putty pads are also available and are often used to seal the back of outlet boxes or other electrical fixtures installed in a membrane penetration. Pads are also used in conjunction with other firestop materials to create firestopping systems.

Fire-rated caulks are easily applied, and installation time is short. However, their shelf life is limited, and precautions may be required when handling some caulking materials. All caulks have definite storage and installation temperature ranges. These materials set up as they cure, and re-entry requires a patch. Also, caulks must be selected based on the type of material to which they will be required to adhere.

Foam compound systems are generally silicone-based and require precision mixing of two components. They are effective on very large openings where dedicated pumping equipment can be used to mix and pump large volumes of foam quickly. Smaller penetrations and patches can be handled using cartridges, which are relatively expensive. Care must be taken to observe installation temperature restrictions when using foam systems. Working time is very short, since foam systems begin to set within three minutes. Complete cure takes about one month.

Cementitious (cement-like) materials consist of a dry powder premixed or mixed with water and are more adaptable to large openings than putty or caulk. When using cementitious materials, it may be necessary to allow for thermal expansion or motion of the penetrating item. Plaster or ordinary grout shall not be used as a

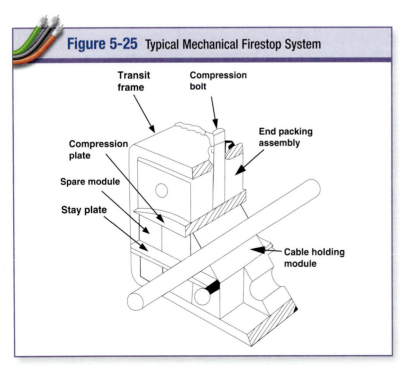

Figure 5-25 Typical Mechanical Firestop System

Transit frame

Compression bolt

Compression plate

End packing assembly

Spare module

Stay plate

Cable holding module

Figure 5-25. Mechanical firestop systems use modules shaped to fit around the cable to provide airtight, chemical-resistant seals. Courtesy of 3M

Figure 5-26 Non-Mechanical Firestop Systems

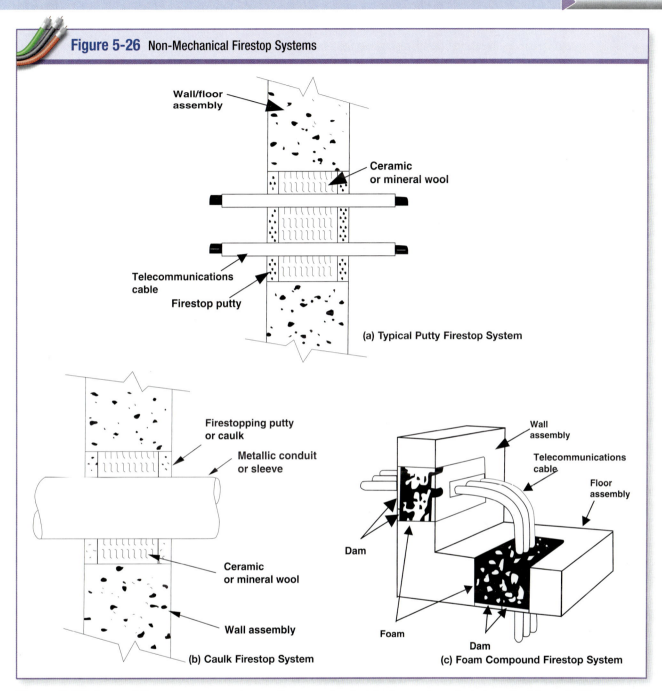

Wall/floor
assembly

Ceramic
or mineral wool

Telecommunications
cable

Firestop putty

(a) Typical Putty Firestop System

Firestopping putty
or caulk

Metallic conduit
or sleeve

Ceramic
or mineral wool

Wall assembly

(b) Caulk Firestop System

Wall
assembly

Telecommunications
cable

Floor
assembly

Dam

Foam

Dam

(c) Foam Compound Firestop System

Figure 5-26. Some of the more common types of non-mechanical firetop systems include putty (a), caulk (b), and foam (c). *Courtesy of 3M*

cementitious firestop. Seals made of grout or plaster may crack, fracture, or fall out, and they may be extremely difficult to re-penetrate.

Intumescent sheets with a sheet metal backing are mostly used for large penetrations in walls and floors. **See Figure 5-27.** Non-sheet-metal backed intumescent sheets may be used with caulk or putty to fabricate a honeycomb-like partitioned

opening for cable, conduit, metal, or non-metallic pipe. When using intumescent sheets, manufacturer's instructions shall be followed for anchoring and sealing.

Intumescent wrap strips or blankets are usually used for firestopping plastic piping, insulated metal piping, cable, cable bundles, nonmetallic conduit, exposed innerduct, or any other material that may burn away in a fire and leave a

Figure 5-27 Intumescent Sheet Firestopping System

Figure 5-27. Intumescent sheets can be used for large penetrations in walls and floors. *Courtesy of 3M*

significant void. **See Figure 5-28.** Intumescent wrap strips may also be used to increase fire endurance. Wrap strips or blankets are relatively soft and permit easy installation around penetrating elements. Intumescent wraps are sometimes used when walls must be built around existing cables.

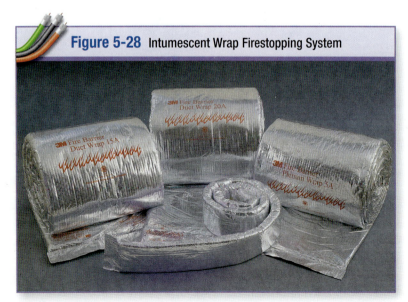

Figure 5-28 Intumescent Wrap Firestopping System

Figure 5-28. Intumescent wrap strips are often used for firestopping flammable materials. *Courtesy of 3M*

Pillows are easy to use, and they work well in areas where a lot of cable activity occurs on a regular basis. **See Figure 5-29.** Pillows are often mistakenly regarded as temporary firestops because of their convenience. Even though pillows are very convenient, pillow firetest qualification standards are the same as for any other sealing system. Earlier pillow designs contained noncombustible fibers; current pillow technology contains a specially treated, compressible fiber matrix. During a fire, this fiber matrix swells and becomes rigid to provide further sealing. The pillow is designed to become rigid so it will withstand the force of water from a hose stream (required by fire test standards).

Note: Some pillows must be restrained by a wire mesh or a metallic sheet for support.

TELECOMMUNICATIONS ROOMS

The telecommunications room(s) on each floor is/are the recognized location(s) of the common access point for backbone and horizontal pathways. The telecommunications room shall be able to contain telecommunications equipment, cable terminations, and associated cross-connect cable/wiring. The telecommunications room shall be located as close as practicable to the center of the area served and should preferably be in the core area. **See Figure 5-30.** All applicable codes shall be observed for the design of the telecommunications room.

Horizontal pathways shall terminate in the telecommunications room located on the same floor as the area being served. Telecommunications room space shall be dedicated to the telecommunications function and related support facilities. Telecommunications room space should not be shared with electrical installations other than those used for telecommunications. Equipment not related to the support of the telecommunications room (for example, water pipes, ductwork, pneumatic tubing, etc.) shall not be installed in, pass through, or enter the telecommunications room.

There shall be a minimum of one telecommunications room per floor. Additional rooms should be provided when the floor area to be served exceeds 10,000 square feet or the horizontal distribution distance to any work area exceeds 295 feet. Telecommunications rooms shall be located on floor areas designed with a minimum floor loading of 50 pounds per square foot. It shall be verified that concentrations of proposed equipment do not exceed the floor loading limit. If unusually heavy equipment is anticipated, these specifications may have to be increased.

A minimum of one wall shall be covered with A/C grade, 19-millimeter (¾-in) plywood (A side exposed) in four-foot by eight-foot sheets hung vertically 150 millimeters (6 in) off the finished floor. The plywood shall be permanently fastened with wall anchors having flat heads giving a flush appearance. It shall be fire-rated or painted with two coats of fire-rated paint prior to any equipment being installed.

The lighting in the telecommunications room shall be a minimum of 500 lux in the horizontal plane and 200 in the verti-

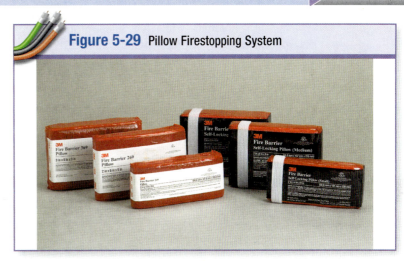

Figure 5-29 Pillow Firestopping System

Figure 5-29. Pillow firestopping is often used where a lot of cabling moves, adds, and changes are anticipated. Courtesy of 3M

cal plane measured one meter (3 ft) above the finished floor. Ideally, the lighting fixtures should be mounted a minimum of 8.5 feet above the finished floor in the middle of the aisle, between cabinets and racks. Lighting fixtures should not be powered from the same electrical distribution panel as the telecommunications

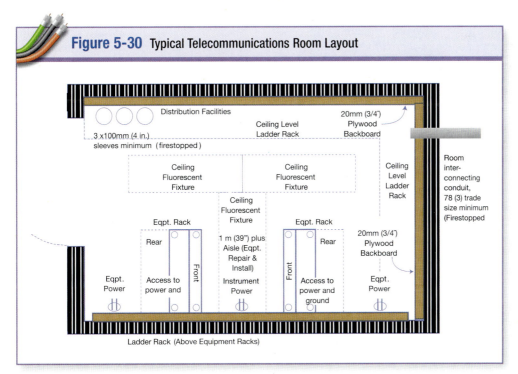

Figure 5-30 Typical Telecommunications Room Layout

Figure 5-30. A typical telecommunications room serves as the common access point for backbone and horizontal pathways.

equipment in the telecommunications room. Dimmer switches should not be used and emergency lighting and signs should be properly placed per the authority having jurisdiction in such a way that an absence of light will not hamper an emergency exit.

For maximum flexibility, a false ceiling shall not be used. The door shall be a minimum of 36 inches wide and 80 inches high, without a doorsill, hinged to open outward (codes permitting), slide side-to-side, or be removable and fitted with a lock. Floors, walls, and ceilings shall be treated to eliminate dust. Finishes shall be light in color to enhance room lighting.

A minimum of two dedicated 120-volt nominal, non-switched, AC duplex electrical outlet receptacles, each on a separate branch circuit, shall be provided for equipment power. These receptacles should be rated at 20 amperes and be connected to a 20-ampere branch circuit.

In addition, identified and marked convenience duplex outlets shall be placed at six-foot intervals around the perimeter walls, at a height of six inches above the floor. If standby power is available, automatic switchover of power should be provided. Specific outlets for equipment and convenience, along with their locations, shall be coordinated with the telecommunications system designers. In many cases, it is desirable that a dedicated power panel be installed to serve the telecommunications room.

Access shall be made available to the telecommunications grounding system specified by TIA-607-C.

Sleeves or slots shall not be left open except during cable installation and shall be properly firestopped per applicable codes.

The telecommunications room is preferably located in a centralized, accessible area on each floor, usually stacked when possible to facilitate cable pulling. Access to shared-use space shall be controlled by the building owner or agent.

Fire protection of the telecommunications room, if required, shall be provided as per applicable code. Sprinkler heads, if required, shall be provided with wire cages to prevent accidental operation.

HVAC shall be included in the design of the telecommunications room to maintain the required temperature and humidity levels. Planning for eventual provisioning, as required, of continuous HVAC (24 hours per day and 365 days per year) shall be included in the initial design; a standalone unit may be used for this purpose. A positive pressure differential shall be maintained with respect to surrounding areas (for dust control) unless prohibited by the authority having jurisdiction. If a standby power source is available in the building, the HVAC system serving the telecommunications room should be connected to the standby supply. Multiple distribution rooms on the same floor shall be interconnected by a minimum of one 3-inch conduit or equivalent pathway.

EQUIPMENT ROOMS

The equipment room is a centralized space for telecommunications equipment, such as a PBX, computer, or LAN server farm that serves occupants of the building. **See Figure 5-31.**

Any or all of the functions of a telecommunications room or entrance facility may alternately be provided by an equipment room.

The room shall house only equipment directly related to the telecommunications system and its environmental support systems. When selecting the equipment room site, avoid locations that are restricted by building components that limit expansion such as elevators, outside walls, or other fixed building walls.

Accessibility for the delivery of large equipment to the equipment room should be provided. Access to shared-use space shall be controlled by the building owner or agent. It is desirable to locate the equipment room close to the backbone pathway.

Floor loading capacity in the equipment room shall be sufficient to bear both the distributed and concentrated load of the installed equipment.

Figure 5-31 Equipment Room

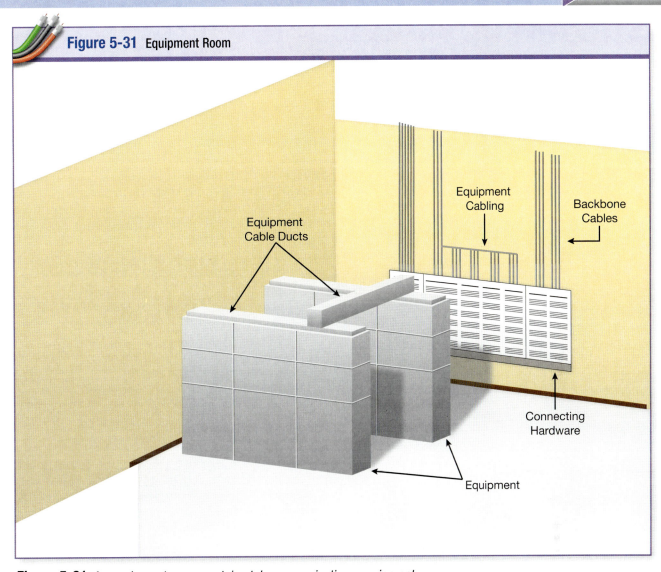

Equipment
Cable Ducts

Equipment
Cabling

Backbone
Cables

Connecting
Hardware

Equipment

Figure 5-31. An equipment room contains telecommunications equipment.

The equipment room shall not be located below water level unless preventive measures against water infiltration are employed. The room shall be free of water or drain pipes not directly required for support of the equipment within the room. A floor drain shall be provided within the room if risk of water ingress exists. The equipment room shall be located with ready access to the main HVAC delivery system. The room shall be located away from sources of electromagnetic interference.

Special attention shall be given to electrical power supply transformers, motors and generators, x-ray equipment, radio or radar transmitters, and induction sealing devices.

When designing the equipment room floor space, allowance shall be made for non-uniform occupancy throughout the building. The practice is to provide 0.75 square feet of equipment room space for every 100 square feet of work area space. The equipment room shall be designed to a minimum of 150 square feet.

Environmental control equipment, such as power distribution or conditioner systems, and uninterruptible power supply (UPS) systems up to 100 kVA shall be permitted to be installed in the equipment room. UPS systems larger than 100 kVA should be located in a separate room. Equipment not related to the support of the equipment room (for example, water pipes, ductwork, pneumatic tubing, etc.) shall not be installed in, pass through, or enter the equipment room.

Minimum clear height in the room shall be eight feet without obstructions. The equipment room shall be protected from contaminants and pollutants that could affect operation and the material integrity of the installed equipment. The equipment room shall be connected to the backbone pathway for cabling to the main terminal space and the telecommunications rooms.

If sprinklers are required within the equipment area, the heads shall be provided with wire cages to prevent accidental operation. Drainage troughs shall be placed under the sprinkler pipes to prevent leakage onto the equipment within the room. For some applications, consideration should be given to the installation of alternate fire-suppression systems.

HVAC shall be provided on a 24 hours-per-day, 365 days-per-year basis. If the building system cannot assure continuous operation for large equipment applications, a standalone unit shall be provided for the equipment room. If a standby power source is available in the building, consideration should be given to connecting the HVAC system serving the telecommunications equipment room to the standby supply. The temperature and humidity shall be controlled to provide continuous operating ranges of 18–27°C (64–81°F) and a relative humidity of 60%. Humidification and dehumidification equipment may be required depending upon local environmental conditions. The ambient temperature and humidity shall be measured at a distance of five feet above the floor level after the equipment is in operation at any point along an equipment aisle center line. A positive pressure differential with respect to surrounding areas should be provided. If batteries are used for backup, adequate ventilation shall be provided. Refer to applicable electrical codes for requirements.

The floor, walls, and ceiling shall be sealed to reduce dust. Finishes shall be light in color to enhance room lighting. Flooring materials having antistatic properties shall be selected.

Lighting shall be a minimum of 500 lux in the horizontal plane and 200 lux in the vertical plane, measured three feet above the finished floor in the middle of all aisles between cabinets. The lighting shall be controlled by one or more switches located near the entrance door(s) to the room. Lighting fixtures should not be powered from the same electrical distribution panel as the telecommunications equipment in the equipment room. Dimmer switches should not be used and emergency lighting and signs should be properly placed in such a way that an absence of light will not hamper an emergency exit.

A separate supply circuit serving the equipment room shall be provided and terminated in its own electrical panel. Electric power provisioning for the equipment room is dependent upon the equipment load and supporting facilities. If a standby power source is available in the building, the equipment room panel should be connected to the standby supply.

The door shall be a minimum of 36 inches wide and 80 inches high, without a doorsill, and shall be fitted with a lock. If it is anticipated that large equipment will be delivered to the equipment room, a double door (72 inches wide by 90 inches high) without a doorsill and center post is recommended. Access shall be made available to the telecommunications grounding system specified by TIA-607-C.

SUMMARY

By following the standards in regards to pathways and spaces, a building owner has greater flexibility to make moves, adds, and changes. Telecommunications infrastructure should be scalable in its design. Telecommunications systems are dynamic and constantly changing. Early infrastructure was designed for voice and data with little thought of interoperability. If the infrastructure is designed correctly in the beginning, then the owner can easily migrate to the latest in communications and integrated building systems technology as they become available.

REVIEW QUESTIONS

1. Non-continuous supports used for cabling shall be located at intervals not to exceed ___?___.
   a. 3'
   b. 5'
   c. 8'
   d. 10'

2. The minimum clear vertical space required above a ceiling tile for the horizontal cabling and pathway is ___?___.
   a. 3"
   b. 5"
   c. 8"
   d. 10"

3. Which floor method consists of a series of channels through which cables pass?
   a. Cellular floor
   b. Raised floor
   c. Underfloor conduit
   d. Underfloor duct

4. When using the sleeve method, sleeves are usually ___?___ in diameter.
   a. 1"
   b. 2"
   c. 3"
   d. 4"

5. When using the slot method, the slot size should be designed at a depth of six inches to ___?___.
   a. 12"
   b. 28"
   c. 24"
   d. 30"

6. What is the minimum amount of access headroom required above a cable tray system or cable runway?
   a. 8"
   b. 12"
   c. 15"
   d. 18"

7. What is the maximum number of feet for a conduit allowed by the standard without the use of a pull box?
   a. 50'
   b. 75'
   c. 100'
   d. 250'

8. What is the maximum amount of conduit fill allowed when only one cable is installed in the conduit?
   a. 23%
   b. 31%
   c. 40%
   d. 53%

9. When using a 2-inch conduit, what is the minimum bend radius of a 90° bend?
   a. 6"
   b. 8"
   c. 10"
   d. 12"

10. The proper size for a splice box with two 1¼-inch conduits is ___?___. (W × L × D)
    a. 4" × 16" × 3"
    b. 12" × 32" × 4"
    c. 14" × 36" × 5"
    d. 18" × 39" × 6"

Cabling System Administration

Introduction

Administration of a telecommunications cabling system includes the proper labeling of all of the elements as well as recordkeeping. The ANSI/TIA-606-C *Administration Standard for Telecommunications Infrastructure* provides a generic procedure to properly label and administer telecommunications infrastructure. Proper identification and labeling of the parts of telecommunications infrastructure reduces maintenance cost and speeds up troubleshooting. Knowing where and what everything is means that less time is required for any moves, adds, or changes (MAC).

Objectives

- Name the standard that addresses telecommunications cabling system administration.
- Name the five areas of telecommunications infrastructure that are administered, according to the standard.
- Cite examples of proper telecommunications infrastructure labeling.
- Explain the requirements of the four classes of the administration standard.

Chapter 6

Table of Contents

ADMINISTRATION STANDARD

The original ANSI/TIA/EIA-606 *Administration Standard for the Telecommunications Infrastructure of Commercial Buildings,* published in 1993, specifies the administrative requirements of telecommunications infrastructure within a new, existing, or renovated commercial building or campus. The newer ANSI/TIA/EIA-606-A *Administration Standard for Telecommunications Infrastructure,* published in May 2002, replaced the 1993 standard and improved upon the original standard because it:

1. Established classes of administration to address the different needs of small, medium, large, and very large telecommunications infrastructure systems
2. Accommodated the scalable needs of telecommunications infrastructure systems
3. Allowed modular implementation of different parts of this standard
4. Specified identifier formats to accommodate the exchange of information between design drawings, test instruments, administration software, and other documents or tools which may be used throughout the lifecycle of the cabling infrastructure
5. Specified labeling formats
6. Harmonized the definitions of terms across all premises' telecommunications infrastructure standards

The TIA-606-B standard replaced the 606-A standard in June 2012, and improved on the 2002 version because it:

1. Adopted the identification scheme specified in TIA-606-A Addendum 1 for racks, cabinets, frames, wall sections, patch panels, and cabling within computer rooms and equipment rooms, and extended the use of these identifiers to locations outside computer rooms and equipment rooms
2. Created a new identification format for Cabling Subsystem 1 links, Cabling Subsystem 2 cables, Cabling Subsystem 3 cables, and campus cabling, but allowed old identifiers for these elements to continue being used
3. Created new identifiers for telecommunications outlets, equipment outlets, splices, consolidation points, and outdoor telecommunications spaces (maintenance holes, pedestals, hand holes, etc.)
4. Extended telecommunications administration to all inter-building telecommunications cabling
5. Permitted administration of Cabling Subsystem 2 and 3 cables by pair groups corresponding to ports rather than copper pairs or single fibers
6. Added administration of telecommunications bonding and grounding systems beyond the telecommunications main grounding busbar (TMGB) and telecommunications grounding busbar (TGB)
7. Permitted existing identifier formats to continue to be used, avoiding the need to create new identifiers and new labels for existing elements
8. Provided information on implementing automated infrastructure management systems

The TIA-606-C standard replaced the 606-B standard in July 2017, and improved on the 2012 version because it:

1. Replaced Addendum 1 to ANSI/TIA-606-B with a reference to ANSI/TIA-5048, the adaption of ISO/IEC 18598 *Information Technology—Automated Infrastructure Management (AIM) Systems—Requirements, Data Exchange, and Applications*
2. Added Annex D with additional guidelines for administration of cabling supporting remote powering, including cable bundle identifiers
3. Removed the preference for the ISO/IEC TR 14763-2-1 compatible format for new administration systems
4. Changed the identifier scheme for telecommunications bonding and

grounding system elements to use ISO/IEC 30129 compatible terms in ANSI/TIA-607-C; legacy terms used in earlier revisions of TIA-607 may also be used:

a. BCT (bonding conductor for telecommunications) changed to TBC (telecommunications bonding conductor)

b. RGB (rack grounding busbar) changed to RBB (rack bonding busbar)

c. GE (grounding equalizer) changed to BBC (backbone bonding conductor)

d. TGB (telecommunications grounding busbar) changed to SBB (secondary bonding busbar)

e. TMGB (telecommunications main grounding busbar) changed to PBB (primary bonding busbar)

5. Added a table with all of the variables used in the identifier formats and renamed some of the variables to avoid duplication

a. L (sequential cable number variable) changed to n

b. s (slot, card, module variable) changed to L

ADMINISTRATION ELEMENTS

Areas of the infrastructure to be administered include:

1. Cabling Subsystem 1 pathways and cabling

2. Cabling Subsystem 2 and 3 pathways and cabling

3. Telecommunications grounding/bonding

4. Spaces (for example, entrance facility, telecommunications room, and equipment room)

5. Firestopping

See Figure 6-1. In addition to providing requirements and guidelines for a traditional paper-based administration system, the ANSI/TIA/EIA-606-C Standard also serves as a platform for the design of computer-based administration

tools which may be necessary as the information base gets larger. The standard does not cover the administration of end-user equipment attached to telecommunications outlets, such as telephone sets and personal computers, nor does it cover application-specific devices in telecommunications spaces, such as switches, LAN hubs, servers, and mainframe computers.

Four classes of administration are specified in this standard to accommodate the diverse degrees of complexity present in telecommunications infrastructure. The specifications for each class include requirements for identifiers, records, and labeling.

CLASSES OF ADMINISTRATION

The standard defines four classes of administration based on complexity of the telecommunications infrastructure.

Class 1 Administration addresses the recordkeeping needs of a premise that is served by a single equipment room (ER). This equipment room is the only telecommunications space (TS) administered; there are no telecommunications rooms (TRs), no Subsystem 2 or 3 cabling, or any outside plant cabling systems to administer. Simple cable pathways will generally be intuitively understood and need not be administered. In order to administer cable pathways or firestopping locations, Class 2 or higher administration should be used. Class 1 Administration will typically be managed using a paper-based system or with general-purpose spreadsheet software.

Class 2 Administration provides for the telecommunications infrastructure administration needs of a single building (or of a tenant) being served by multiple telecommunications spaces (for example, an ER with one or more TRs) within a single building. Class 2 Administration includes all elements of Class 1 Administration, as well as identifiers for Subsystem 2 and 3 cabling, all elements of the grounding and bonding systems, and firestopping. Cable pathways may be

Figure 6-1 Administrative Elements

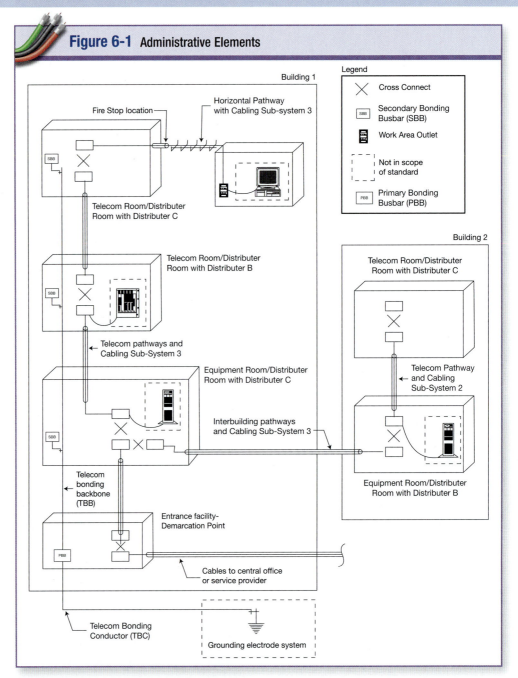

Figure 6-1. *Administrative elements include various pathways and cabling, grounding and bonding systems, spaces, and firestopping.*

intuitively understood, so administration of these elements is optional. Class 2 may be managed using a paper-based system, general-purpose spreadsheet software, specialized software, or an automated infrastructure management (AIM) system.

Class 3 Administration addresses the needs of a campus environment, includ-

ing all of the buildings and outside plant elements. Class 3 Administration includes all elements of Class 2 Administration, along with identifiers for all buildings and inter-building campus cabling. Administration of building pathways and spaces and outside plant elements is recommended. Class 3 may be managed with a paper-based system, general-purpose spreadsheet software, specialized software, or an AIM system.

Class 4 Administration addresses the needs of a multi-campus/multi-site system. Class 4 Administration includes all elements of Class 3 Administration, plus an identifier for each site and optional identifiers for any intercampus elements (for example, wide area network connections). For mission critical systems, large buildings, or multi-tenant buildings, administration of pathways and spaces and outside plant elements is strongly recommended. Class 4 may be managed with general-purpose spreadsheet software, specialized software, or an AIM system.

Class 1 Administration

Class 1 administration requires infrastructure identifiers for the following elements:

- Telecommunications space (TS)
- Cabinets, racks, enclosures, and wall segments
- Patch panels or termination blocks
- Patch panel port and termination block position
- Cables between cabinets, racks, enclosures, or walls in the same space
- Cabling subsystem 1 (horizontal) links
- Primary bonding busbars (PBB)
- Secondary bonding busbars (SBB)

Telecommunications Space Identifier

The TIA-606-C standard requires a telecommunications space identifier to be assigned to the space. This identifier must be unique within the building.

The TIA-606-C standard requires the format of the TS identifier to be:

$$fs$$

f numeric character(s) identifying the floor of the building occupied by the TS. In buildings with only a single floor, this identifier is optional.

s alpha character(s) uniquely identifying the TS or computer room on floor f.

The TIA-606-C standard allows alphanumeric characters to be used in the "f"

format for buildings with non-numeric floors. The standard requires the naming convention to be consistent within the building and also suggests that all TS identifiers in a single infrastructure should have the same format.

For example, 3TR would be the identifier for a Telecom Room on the 3rd floor of a building. Room 405 may simply have the identifier 405.

The TIA-606-C standard requires the TS to be labeled with the TS identifier inside the room in such a way as to be visible to someone working in that room.

ISO/IEC Compatible Format

The TIA-606-C standard also lists the ISO/IEC TR 14763-12-1 compatible format. In most instances the compatible format simply adds a "+" in front of the labeling format.

For example, the TS identifier format compatible with ISO/IEC TR 14763-2-1 is: +fs

*Note that this is similar to the TIA-606-C compatible format, but with a "+" indicating a location aspect of a building instead of an object.

For example: +3TR would be the ISO/IEC compatible identifier for a Telecom Room on the 3rd floor of a building. Room 405 would have the identifier +405.

More ISO/IEC compatible labeling formats are available throughout the TIA-606-C standard.

Cabinet, Rack, and Wall Segment Identifier

Identifiers are needed in telecomminications spaces for all cabinets, racks, and wall spaces housing telecommunications cabling or equipment. In high-density areas such as data centers and computer rooms, there must be a reliable way to label all components for easy location and administration.

Cabinet and Rack Identifiers Where Grid Coordinates are Used. The TIA-606-C standard recommends using a grid coordinate system in a telecommunications space with multiple rows of cabinets or racks, such as computer rooms and equipment rooms, to identify the equipment cabinets and racks located within the room. In rooms that have access floor systems, the standard requires use of the access floor grid identification scheme to be used for this purpose. If the room does not have an access floor system, the ceiling tile grid should be used as the basis for identification. In rooms that have neither a floor tile grid nor a ceiling tile grid, the standard suggests that a grid be applied to the floor plan.

A grid applied to a floor plan must be dense enough to ensure that no two cabinets occupy the same grid space. The standard suggests grid spacing between 500 millimeters and 600 millimeters (20-24 in).

The standard suggests using the long axis of the room as the "X" axis and the short axis of the room as the "Y" axis. **See Figure 6-2.** The characters used to label the "X" and "Y" axes must be adequate to cover the entire space covered by the grid. The standard does not have any requirements on where the starting point needs to be in the grid system, but if there is a possibility that the space may be expanded in the future, the standard suggests setting the starting point of the grid in a corner of the room that is unlikely to expand.

The TIA-606-C standard recommends marking the grid coordinates on the walls of the room. If floor or ceiling tiles are used as the basis for a grid system, they may also be marked.

The TIA-606-C standard requires the grid coordinate location identifier to have a format of:

$$fs.xy$$

fs   The telecommunications space or computer room identifier.

x   One or more alphabetic characters designating the "X" coordinate of the rack or cabinet. The number of characters used for the "X" coordinates must be the same throughout the entire space, so if the space requires between 27 and 676 coordinates along the "X" axis, the standard requires the "X" axis sequence to start with "AA" rather than "A."

y   One or more numeric characters designating the "Y" coordinate of the rack or cabinet. The number of digits used for the "y" coordinate must be the same throughout the entire space covered by the grid, so the standard requires the "Y" axis to start with "00" or "01" rather than "0" or "1" when the number of coordinates is more than 9 but fewer than 100.

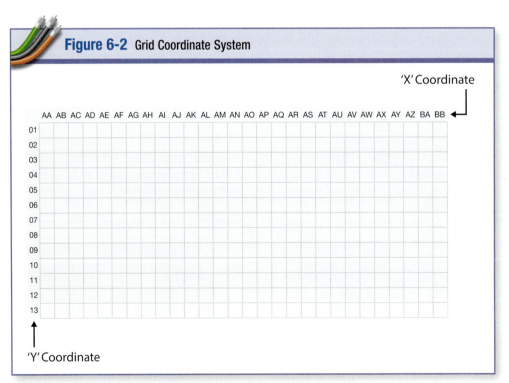

Figure 6-2 Grid Coordinate System

Figure 6-2. The grid system consists of an X and Y axis.

It is possible for cabinets and racks to occupy more than one square in rooms using the grid system. In this scenario, the standard requires the same method to be used to determine the grid location for every cabinet or rack. The location of a given rack may be determined to be the grid space containing the corner of the rack closest to the starting point of the grid, the grid space containing the left front corner, the grid space containing the right front corner, or the grid space containing the front center, as long as the same location is used throughout the room. For example, if the identifier of a cabinet is determined by the floor space grid containing the right front corner of the cabinet, the cabinet with its right front corner on tile AD02 will have the identifier AD02. **See Figure 6-3**.

Using this convention will allow for cabinets and frames to be replaced with different sized equipment without changing the identification of existing equipment.

The standard requires wall-mounted systems in rooms with a grid system to use the grid coordinates of the wall space. It recommends that the wall space be divided into sections the width of each grid coordinate. For example, if a wall-mounted rack is mounted above tile AD01, it would have an identifier of AD01.

Cabinet and Rack Identifiers Where Grid Coordinates are Not Used. For rooms that do not have grid coordinates, the standard allows cabinets and racks to be identified by their row number and their location within the row. This method is only recommended if the space meets the following criteria:

- There are only a small number of cabinets in a few rows, or
- There are uniformly spaced or static equipment rows (rows that will not be reoriented or replaced with more or fewer rows), and
- The cabinets, racks, and frames have uniform widths and will not be replaced with those of different widths.

The standard does require the quantity of characters used to be the same throughout the space.

When grid coordinates are not available, the standard requires the identifier to have a format of:

$$fs.xy$$

fs The telecommunications space or computer room identifier.

x One or more characters designating the row containing the cabinet or frame. The quantity of characters used for the row identifier should be the same throughout the entire space. When there are more than nine rows, the standard recommends that these characters be alphabetic rather than numeric. If there is only one row in the telecommunications space, this character is optional.

y One or more characters designating the location of the cabinet or frame within the row. The quantity of characters used should be the same throughout the space. In addition, the standard recommends that the

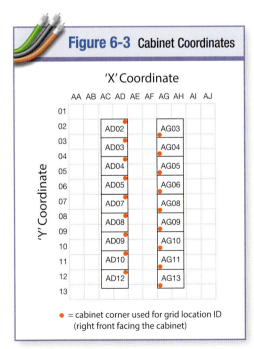

Figure 6-3 Cabinet Coordinates

'X' Coordinate

● = cabinet corner used for grid location ID (right front facing the cabinet)

Figure 6-3. Grid systems may have cabinets that occupy more than one grid.

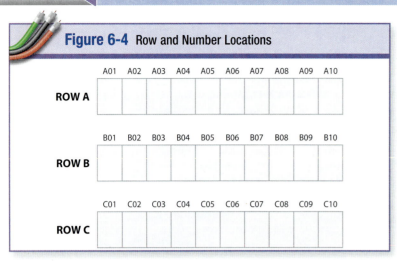

Figure 6-4 Row and Number Locations

Figure 6-4. The standard allows for row number/location within the row designations to be used when grid coordinates are not.

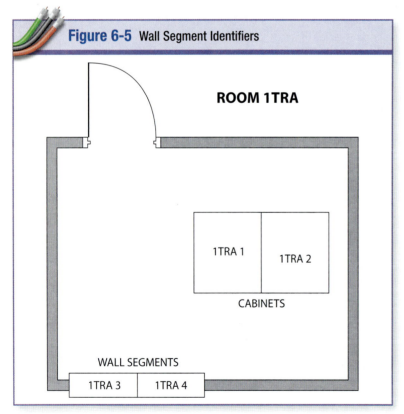

Figure 6-5 Wall Segment Identifiers

ROOM 1TRA

1TRA 1   1TRA 2

CABINETS

WALL SEGMENTS

1TRA 3   1TRA 4

Figure 6-5. The standard recommends dividing a wall space into sections when a grid is not used.

location identifiers be consistent between rows, with numbers starting from the same end of each row and increasing in the same direction.

For locations that do not use the grid identifier system, the standard recommends that each row be labeled with its row identifier at both ends of the row. It also recommends that the identifiers be sequential and unique, and mandates that the same format be used for all rows within the room. **See Figure 6-4.**

The standard also recommends identifiers for each wall where telecommunications equipment is mounted in a room without a grid system. It suggests dividing the wall space into sections that correspond to the frame or cabinet numbers. The sections can be the width of a typical cabinet or frame, or the wall can be divded into one-meter wide sections and labeled starting from the left edge of the wall. **See Figure 6-5.**

Cabinet and Rack Labeling. The TIA-606-C standard requires all racks and cabinets to be labeled on the front and the back of the rack or cabinet in plain view. Though it is not required, the standard recommends that the labels be located on the top or bottom on a permanent part of the cabinet or rack. **See Figure 6-6.**

Patch Panel and Termination Block Identifiers

The TIA-606-C standard requires patch panels mounted in cabinets and racks in a single vertical column to have identifiers using the format:

$$fs.x1y1\text{-}r1$$

$fs.x1y1$ The cabinet, rack, frame, or wall segment identifier.

$r1$ The standard identifies three options that are permitted formats for $r1$:

Option 1: $r1$ = Two numerical digits designating the location of the top of the patch panel in rack units (U) from the bottom of the usable space in the cabinet or frame. This is the format recommended in the standard.

Option 2: $r1$ = One letter indicating the side of the cabinet or frame, followed by two numerical digits designating the location of the top of the patch panel in rack units from the bottom of the usable space in the cabinet or frame. The letters used to indicate the sides of the cabinet or frame can be any unique set of letters as long as they are consistently used within the infrastructure. For example, A, B, C, and D could be used for the four sides of a cabinet starting from the front and proceeding clockwise (when viewed from the top), or N, S, E, and W could be used for the four sides of the cabinet if the sides are aligned with the four compass directions. In another example, if only the front and rear of a cabinet are used, they could be labeled F and R respectively.

Option 3: $r1$ = One to two characters designating the patch panel location within the cabinet or rack, beginning at the top. The standard does not include horizontal cable managers when sequencing patch panels. However, it does require the quantity of characters to be the same for all patch panels in the cable or rack. If $r1$ uses a mixture of alphabetical and numeric characters, the letters "I," "O," and "Q" must be excluded.

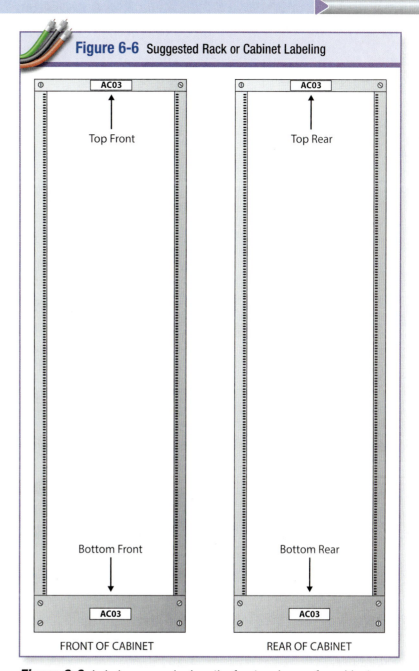

Figure 6-6 Suggested Rack or Cabinet Labeling

AC03
Top Front
Bottom Front
AC03
FRONT OF CABINET

AC03
Top Rear
Bottom Rear
AC03
REAR OF CABINET

Figure 6-6. *Labels are required on the front and rear of a cabinet or rack.*

The standard requires all patch panels to be labeled with their identifiers. The room name, however, is not required.

The standard requires that each port, the first port, or the last port of each subpanel be labeled. For example, if a patch panel has an identifier of 1DC-

AD02-31, it indicates that the panel is located in room 1DC, cabinet AD02, and is the 31st patch panel, counting up from the bottom of the rack.

Although the patch panels occupy multiple rack unit positions, each patch panel is identified by the rack unit

Figure 6-7 Patch Panel Identification

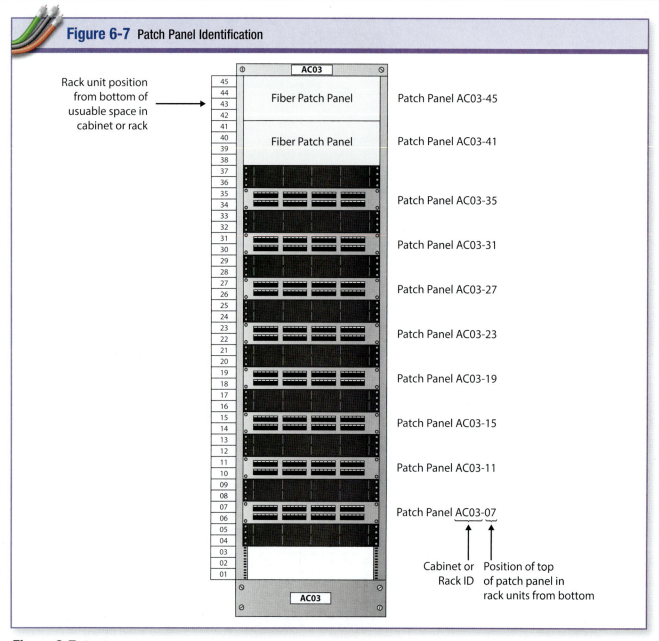

Rack unit position from bottom of usable space in cabinet or rack

AC03

Fiber Patch Panel — Patch Panel AC03-45

Fiber Patch Panel — Patch Panel AC03-41

Patch Panel AC03-35

Patch Panel AC03-31

Patch Panel AC03-27

Patch Panel AC03-23

Patch Panel AC03-19

Patch Panel AC03-15

Patch Panel AC03-11

Patch Panel AC03-07

AC03

Cabinet or Rack ID   Position of top of patch panel in rack units from bottom

Figure 6-7. *Rack position labels start from the bottom of the rack.*

position of the top of the patch panel. **See Figure 6-7.**

The ports on the patch panel should be labeled where space is available. When a patch panel supports Cabling Subsystem 2 or 3, the standard suggests that the labeling indicate the name of the space to which the cables run.

The TIA-606-C standard requires all ports on patch panels and all positions on termination blocks to be labeled with the corresponding port number or position number. Additional identifier fields may be included as practicable. All subpanels are required to be labeled with their subpanel identifier.

Balanced Twisted-Pair Patch Panels

Consider a 48-port balanced twisted-pair patch panel at 35U from the bottom of cabinet AD02, with identifier AD02-35 with:
- 12 UTP cables to the patch panel 35U from the bottom of cabinet AG03, ports 01-12
- 12 UTP cables to the patch panel 31U from the bottom of cabinet AG04, ports 01-12
- 12 UTP cables to the patch panel 45U from the bottom of cabinet AG05, ports 01-12
- 12 UTP cables to the patch panel 42U from the bottom of cabinet AG06, ports 01-12

The labels below each group of six ports include the local and far-end patch panel and port identifiers. **See Figure 6-8.**

Now consider a patch panel 35U from the bottom of cabinet AD03 (patch panel AD03- 35) with three individual cables with 24 strands of multimode fibers (12 pairs of multimode fibers) in each to:
- Patch Panel AG10-41, ports 01-12
- Patch Panel AG11-41, ports 01-12
- Patch Panel AG13-41, ports 01-12

See Figure 6-9. Fiber patch panel labels may include additional information such as cable type, near-end TS name, and far-end TS name.

The first line on the cover label identifies where the patch panel is located, that it terminates multi-mode fiber, and that the patch panel ID is AD03-45. The following lines of the cover label specify the patch panel IDs to which the cables on each port terminate.

Figure 6-8 Port Identifiers for Cabling Subsystems 2 or 3

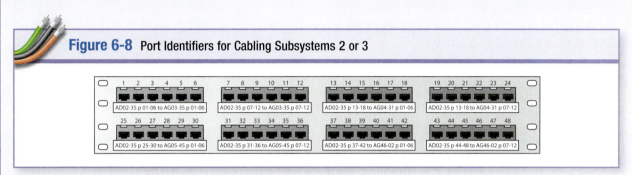

Figure 6-8. When space permits, labels should indicate the far-end ports.

Figure 6-9 Labeling Fiber Patch Panels

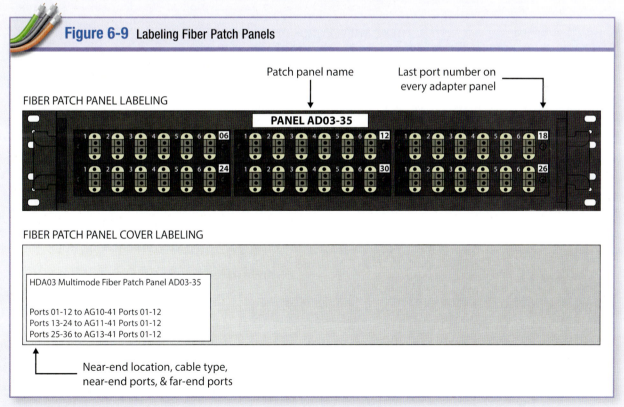

Figure 6-9. Fiber patch panel labels should include location information for the near- and far-end ports and the cable type.

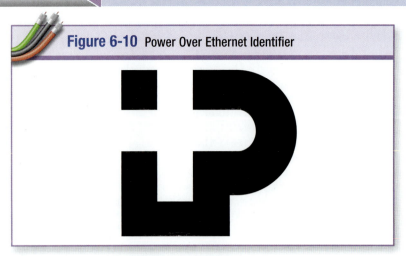

Figure 6-10 Power Over Ethernet Identifier

Figure 6-10. *The TIA-606-C standard identifies the standard symbol for Power over Ethernet (POE) applications.*

Where applications provide electrical power in addition to data transmission over balanced twisted-pair cables, visual segregation and identification of ports with power can be accomplished through the use of a symbol. **See Figure 6-10.**

Cables between Patch Panels or Termination Blocks

The TIA-606-C standard requires all cables terminated on patch panels or termination blocks to be identified with the identifiers of the ports/terminations on both ends of the cable separated by a forward slash. If the cable supports multiple ports/termination positions, then the standard recommends that the first and last port/termination on each end of the cable be provided in the identifier.

Cabling Subsystem 1 Link Identifier

The TIA-606-C standard requires a Cabling Subsystem 1 (horizontal cable) link identifier, unique within the administration system, to be assigned to each Cabling Subsystem 1 link and to its elements.

Cabling Subsystem 1 links terminated on an equipment outlet are required to use the format:

$$fs\text{-}ak$$

fs   The TS identifier for the location of the patch panel or termination block on which the cable terminates. The standard does not require this portion of the identifier when the Class 1 Administration system is limited to a single equipment or computer room.

a   One or two alpha characters uniquely identifying a single patch panel, a group of patch panels with sequentially numbered ports, a termination block, or a group of termination blocks, serving as part of the horizontal cross-connect.

k   Two to four numeric characters designating the port on a patch panel or the section of a termination block on which a Cabling Subsystem 1 link is terminated in the TS. The standard requires enough numeric characters to be used for this portion of the identifier to accommodate all Cabling Subsystem 1 links in a distributor.

The elements identified by the standard for a balanced twisted-pair Cabling Subsystem 1 link include:

1. The connecting hardware, e.g., patch panel port or the position of a termination block terminating a 4-pair Cabling Subsystem 1 cable
2. A 4-pair Cabling Subsystem 1 cable
3. An equipment outlet terminating a 4-pair Cabling Subsystem 1 cable in the equipment outlet space
4. If a consolidation point (CP) is present:
   a. The segment of 4-pair Cabling Subsystem 1 cable extending from the TS to the CP connecting hardware
   b. The CP connecting hardware or section of an IDC connector terminating a 4-pair Cabling Subsystem 1 cable
   c. The segment of 4-pair Cabling Subsystem 1 cable extending from the CP connecting hard-

---

FACT

For a single-dwelling residence, the standard recommends including the room and wall location of each outlet on the patch panel if the outlets are not labeled. For example, B1-N could be used as the label to denote an outlet on the north wall of Bedroom 1.

If single-dwelling residential outlets are not labeled, the standard suggests the inclusion of simple floor plans at the distribution device that provide a graphic correlation between physical outlet locations and outlet designations.

ware to the outlet/connector, if present

5. If a MUTOA is present, telecommunications outlet/connector in the MUTOA

Note: The elements identified by the standard for a fiber Cabling Subsystem 1 link are similar to those of the Cabling Subsystem 1.

For the pair of fiber terminations, the standard allows either two simplex connectors or one duplex connector, and includes adapters, if present.

All patch panels and termination blocks are required to be labeled. When the *fs-ak* format is used for Cabling Subsystem 1 link identifiers, each patch panel port or position of a termination block in the TS is required to be labeled with the *ak* portion of the identifier.

The standard requires each end of a Cabling Subsystem 1 cable to be labeled within 300 millimeters (12 in) of the end of the cable jacket with the Cabling Subsystem 1 link identifier, which must be visible on the exposed part of the cable jacket. This includes each cable end in the TS, at the equipment outlet space, and at the CP, if present.

In commercial buildings, industrial premises, data centers, and multi-tenant buildings, each individual telecommunications outlet or equipment outlet is required to be labeled with the Cabling Subsystem 1 link identifier. The labeling must appear on the connector, faceplate, or MUTOA in a way that clearly identifies the individual connector associated with the particular identifier. In single-dwelling residences, labeling of telecommunications outlets is recommended, but not required.

Primary Bonding Busbar Identifier

The standard requires a primary bonding busbar (PBB) identifier.

The PBB identifier is used to identify the single PBB present in a building. (For more information on grounding and bonding and the products used, see Chapter 7.)

$$fs\text{-PBB}$$

fs   The identifier for the space containing the PBB.

-PBB   Designates the element as being a primary bonding busbar.
Note: In instances where the legacy term "telecommunications main grounding busbar" is used, the standard allows -PBB to be replaced with -TMGB.

The standard requires the PBB to be labeled on the front with the PBB identifier. When it is not practical to label the surface of the PBB, then the label is required to be on the wall near the PBB.

Secondary Bonding Busbar Identifier

The standard requires a secondary bonding busbar (SBB) identifier.

The SBB identifier is used to identify SBBs in the bonding and grounding system. The TIA-606 standard requires a unique SBB identifier to be assigned to each SBB.

$$fs\text{-SBB}[i]$$

fs   The identifier for the space containing the SBB.

-SBB   Designates the element as being a secondary bonding busbar.
Note: If the legacy term "telecommunications grounding busbar" is used, then the standard allows -SBB to be replaced with -TGB.

i   The optional sequence number, starting at "1," to be used if there is more than one SBB in the space.

The TIA-606-C standard requires each SBB to be labeled on the front with the SBB identifier. Where it is not practical to apply the label on the front surface of the SBB, the identifier can be placed on a nearby wall.

Rack Bonding Busbar Identifier

The TIA-606-C standard provides information about administration for a rack bonding busbar (RBB). The RBB identifier is used to identify RBBs in racks, cabinets, and frames. The RBB identifier is an optional identifier and is not required per the TIA-606-C standard; however, if the RBB identifier is used, the format is:

$$fs.xy\text{=RBB}[j]$$

fs.xy   The identifier of the cabinet, rack, or frame.

=RBB   Designates the element as being a rack bonding busbar.
Note: If the legacy term "rack grounding busbar" (RGB) is used, then =RBB may be replaced with =RGB.

j   The optional sequence number, starting at "1," to be used if there is more than one RBB in the cabinet, rack, frame, or wall segment.

Where the optional RBB identifier is used, it should be labeled with its identifier in clear view of anyone making a termination on the RBB.

Mesh Bonding Network Identifier

The TIA-606-C standard provides information about administration for a mesh bonding network identifier (mesh-BN). This is used to identify common networks in a computer room, equipment room, or other space. The mesh-BN identifier is optional and is not required per the TIA-606-C standard. However, if the mesh-BN identifier is used, the format is:

$$fs=MBN$$

fs   The identifier for the space containing the mesh-BN.

=MBN   Designates the element as being a mesh-BN.

When a mesh-BN identifier is used, there is no requirement for labeling the actual mesh-BN, but all connections made to the mesh-BN should be labeled.

Telecommunications Bonding Conductor Identifier

The TIA-606-C standard requires a telecommunications bonding conductor (TBC) identifier. The TBC identifier is used to identify the telecommunications bonding and grounding system and is required to be unique.

The format for TBC identifiers is:

$$fMsM / fEsE=TBC$$

fMsM   The TS identifier for the space containing the PBB.

fEsE   The TS identifier for the space containing the TBC. This is typically the electrical entrance facility that contains the service

equipment (power) ground to which the telecommunications bonding conductor is attached.

=TBC   Designates the element as being the telecommunications bonding conductor.
Note: If the legacy term "bonding conductor for telecommunications" is used, the standard allows =BCT to be used here instead.

The standard requires the TBC to be labeled with its identifier on both ends. The labels must be durably affixed to both ends and conspicuously displayed just prior to the conductor being routed into its termination.

Telecommunications Bonding Backbone Identifier

The TIA-606-C standard requires a telecommunications bonding backbone (TBB) identifier. The TBB identifier is used to identify a TBB in the bonding and grounding system. The TBB is identified by the names of the secondary bonding busbars on either end of the TBB.

The formats for TBB identifiers are:

$$fMsM\text{-}PBB / f2s2\text{-}SBB[i2]$$
$$\text{or}$$
$$f1s1\text{-}SBB[i1] / f2s2\text{-}SBB[i2]$$

fMsM-PBB   The identifier for the PBB.

f1s1-SBB[i1] and *f2s2-SBB[i2]*   The identifiers for the SBBs.

Each TBB must be labeled with its identifier on both ends. The labels are required to be durably affixed to both ends and conspicuously displayed just prior to the conductor being routed into its termination.

Backbone Bonding Conductor Identifier

The TIA-606-C standard requires a backbone bonding conductor (BBC) identifier. The BBC identifier is used to identify a BBC in the bonding and grounding system. The standard recommends that the identifier be unique and have the format based on the identifiers of the SBBs on either end of the BBC separated by a forward slash ("/"). Each BBC is required to be labeled on both ends.

The format for BBC identifiers is:

$f1s1\text{-}SBB[i1] / f2s2\text{-}SBB[i2]$

Identifier for Bonding Conductors Attached to a Primary Bonding Busbar

The TIA-606-C standard requires all bonding conductors attached to a PBB to have a unique identifier. The format is:

$fs\text{-}PBB / object$

fs-PBB   The identifier of the PBB.

$object$ The identifier of the object to which the bonding conductor is attached. This could be the identifier of a cabinet/rack, a mesh-BN, an RBB, an electrical panel, a pathway, building steel, a cable tray system, or equipment such as a LAN switch.

The standard requires all conductors attached to the PBB to be labeled on both ends.

Identifier for Bonding Conductors Attached to a Secondary Bonding Busbar

The TIA-606-C standard requires all bonding conductors attached to an SBB to have a unique identifier. The format is:

$fs\text{-}SBB / object$

fs-SBB   The identifier of the SBB.

$object$ The identifier of an object to which the bonding conductor is attached. This object could be the identifier of a cabinet/rack, a TBB, a mesh-BN, an RBB, an electrical panel, a cable tray system, or equipment such as a LAN switch.

The standard requires all bonding conductors attached to an SBB to be labeled at both ends.

Note: The standard suggests the same format be used for objects connecting to a mesh bonding network or rack bonding busbar using the applicable acronym for each.

Required Records

The TIA-606-C Standard provides the requirements for Class 1 Administration records.

The standard requires one Cabling Subsystem 1 link record for each Cabling Subsystem 1 link. Cabling Subsystem 1 link records must contain the following information:

1. The Cabling Subsystem 1 link identifier
2. The cable type (e.g., 4-pair, UTP, category 6, plenum)
3. The location of telecommunications outlet/connector
4. The outlet connector type (e.g., 8-position modular, category 6)
5. The cable length
6. The cross-connect hardware type (e.g., 48-port modular patch panel)
7. The service record of the link (for example: "passed category 6 at installation 1/12/11, re-terminated and re-tested at cross-connect 4/22/11 due to broken wire").

The standard allows additional items of information desired by the system owner or operator to be added at the end of the record. The added information could be information such as the location of test results, the location of the outlet within the room or office, or other telecommunications outlet/connectors at same location (generally, the other outlet connectors in the same faceplate).

Class 2 Administration

Class 2 Administration addresses infrastructure with one or more telecommunications spaces in a single building.

Class 2 Administration requires infrastructure identifiers for the following elements:

- Identifiers required in Class 1 Administration
- Building Cabling Subsystem 2 and 3 (backbone) cables
- Building Cabling Subsystem 2 and 3 ports
- Firestopping locations

The TIA-606-C standard also allows for optional pathway identifiers used in Class 2 Administration. These optional identifiers are identified in clause 9 and include:

- Outdoor telecommunications space identifiers

- Intra-space pathway identifiers
- Building pathway identifiers
- Building entrance pathway identifiers
- Outside plant pathway identifiers
- Campus entrance pathway identifiers

Building Cabling Subsystem 2 and 3 Cable Identifiers

The TIA-606-C standard requires a Cabling Subsystem 2 or 3 cable identifier that is unique within the building to be assigned to each Cabling Subsystem 2 and 3 cable. The standard provides two options for this identifier:

Option 1:

$$f1s1.x1y1\text{-}r1\text{:}P1[\text{-}P2] / f2s2.x2y2\text{-}r2\text{:}P3[\text{-}P4]$$

The following example explains the format:

1TER.AD04-40:1-24/3TRC.2-45:1-24

f_1s_1 x_1y_1 r_1 $P_1\text{-}P_2$ f_2s_2 x_2y_2 r_2 $P_3\text{-}P_4$

$f1s1$	1TER	1st floor Telecom Entrance Room
$x1y1$	AD04	Cabinet at grid coordinates AD04
$r1$	-40	Patch panel at rack unit 40 (RU 40)
$P1\text{-}P2$	1-24	Ports 1-24
$f2s2$	3TRC	3rd floor Telecom Room C
$x2y2$	2	Rack 2
$r2$	45	Patch panel at rack unit 45 (RU 45)
$P3\text{-}P4$	1-24	Ports 1-24

Option 2:

$$f1s1 / f2s2\text{-}n$$

$f1s1$	The TS identifier for the space containing the termination of one end of the Cabling Subsystem 2 or 3 cable.
$f2s2$	The TS identifier for the space containing the termination of the other end of the Cabling Subsystem 2 or 3 cable.
n	One or two alphanumeric characters identifying a single cable

with one end terminated in the TS designated $f1s1$ and the other end terminated in the TS designated $f2s2$.

The $f1s1 / f2s2\text{-}n$ format is allowed for compatibility with administration systems that use previous revisions of the TIA-606 Standard.

Cabling Subsystem 2 and 3 labels must be placed on both ends of each cable within 12 inches of the end of the cable jacket.

Pairs, Strands, and Grouping Identifiers for Building Cabling Subsystem 2 and 3

The TIA-606-C standard requires each pair or fiber strand grouping corresponding to a port on a building Cabling Subsystem 2 or 3 cable to have a unique identifier. The standard gives two compatible formats for pairs or ports on building Cabling Subsystem 2 and 3 cables. The first is:

$$f1s1.x1y1\text{-}r1\text{:}P1 / f2s2.x2y2\text{-}r2\text{:}P2$$

P1 and P2 designate the pairs or ports used for the cabling subsystem.

The second identifier format for pairs or ports on building Cabling Subsystem 2 and 3 cables is:

$$f1s1 / f2s2\text{-}n.d$$

$f1s1/$ $f2s2\text{-}n$	The Cabling Subsystem 2 or 3 cable identifier
d	Two to four numeric characters identifying a balanced twisted-pair group, fiber, or grouping corresponding to a port.

When the second identifier format is used, it must include the 'd' as that is what defines the type of pairs in the subsystem.

Since individual optical fibers and balanced pairs are typically color-coded rather than individually labeled, the standard does not require labeling. However, break-out cables with one connector on one end and multiple connectors on the other are required to be labeled.

Firestopping Location Identifier

The TIA-606-C standard requires a firestopping location identifier for each installation of firestopping material. The standard suggests that all firestopping location identifiers in a single infrastructure should have the same format where possible.

The legacy format for firestopping location identifiers is:

$$f\text{-FSL}w(h)$$

f Numeric character(s) identifying the floor of the building occupied by the TS or computer room.

-FSL Identifies the element as being a firestopping location.

w Two to four numeric characters identifying one firestopping location.

h One numeric character specifying the hour rating of the firestopping system.

The TIA-606-C standard requires each firestopping location to be labeled at each location where firestopping is installed, on each side of the penetrated fire barrier within 12 inches of the firestopping material.

Note: The new format defines pathways and distance from the end of the pathway and is much more detailed than likely necessary.

Required Records
The standard contains the requirements for records required in Class 2 Administration. The required records include:

1. Cabling Subsystem 1 link records
2. One TS record for each TS
3. One Cabling Subsystem 2 or 3 cable record for each Cabling Subsystem 2 or 3 cable
4. One PBB record
5. One SBB record for each SBB
6. One firestopping location record for each firestopping location.

TS records. The TS records are required to contain the following information:

1. The TS identifier
2. The type of TS (e.g., TR, ER, or EF)
3. The building room number
4. The key or access card identification
5. The contact person
6. The hours of access

Building Cabling Subsystem 2 and 3 cable records. The building Cabling Subsystem 2 and 3 cable records are required to contain the following information:

1. The building Cabling Subsystem 2 and 3 cable identifiers
2. The type of cable

3. The type of connecting hardware (first TS)
4. The type of connecting hardware (second TS)
5. A cross-connect table relating each Cabling Subsystem 2 and 3 cable pair or fiber

PBB records. The primary bonding busbar records are required to contain the following information:

1. The PBB identifier
2. The location of the PBB
3. The location of attachment of the PBB to electrical system ground or building structural steel
4. The location of test results for any tests performed on the PBB, such as resistance to ground

SBB records. The secondary bonding busbar records are required to contain the following information:

1. The SBB identifier
2. The location of the SBB
3. The location of test results for any tests performed on the SBB, such as resistance to ground

Firestopping records. The firestopping records are required to contain the following information:

1. Firestopping location identifier
2. Location of the firestopping installation
3. Type and manufacturer of firestopping installed
4. Date of firestopping installation
5. Name of installer of firestopping material

Class 3 Administration
Class 3 administration addresses infrastructure with multiple buildings at a single site.

Infrastructure Identifiers
The TIA-606-C standard requires the following infrastructure identifiers in Class 3 Administration:

- All identifiers required in Class 2 Administration
- Building identifiers
- Campus cable identifiers
- Campus cable pair or fiber identifiers

The TIA-606-C standard also permits the following optional infrastructure identifiers in Class 3 Administration:

- Identifiers optional in Class 2 Administration
- Outside plant pathway element identifiers
- Campus pathway or element identifiers

Building Identifier

The TIA-606-C standard requires a unique building identifier to be assigned to each building on the site. The TIA-606-A compatible format for building identifiers is:

$$c$$

c One or more alphanumeric characters identifying a building on a site.

Required Records

Required records for Class 3 Administration include:

1. All records required in Class 2 Administration
2. One building record for each of the buildings on the site including name, location, all TSs, contact information, and access hours
3. One inter-building campus cable record for each campus cable that includes cable identifiers, the type of cable, connecting hardware, and cross-connections

Class 4 Administration

Class 4 Administration addresses infrastructure with multiple sites or campuses.

Permanent Labels

The TIA-606-C standard addresses the infrastructure identifiers required in Class 4 Administration. The Class 4 infrastructure identifiers required are:

- All identifiers required in Class 3 Administration
- Campus or site identifiers

Optional infrastructure identifiers in Class 4 Administration include:

- Identifiers optional in Class 3 Administration
- Inter-campus element identifiers

Required Records

Required records for Class 4 Administration include:

1. All records required in Class 3 Administration
2. One campus or site record for each campus or site

Campus or site records are required to contain:

1. The campus or site name
2. The campus or site location (e.g., street address)
3. The contact information for local administrator of infrastructure
4. A list of all buildings at the site or campus
5. The location of the main cross-connect, if applicable
6. The access hours

The TIA-606-C standard requires labels for all administration tasks to be:

- Machine printed
- Uppercase text
- A font without serifs
- Large enough to be easily read
- Resistant to the environment in which they are located
- High contrast to be easily read
- Durably affixed to the cable on both ends
- Conspicuously displayed just prior to each cable being routed into the termination device

The TIA-606-C standard is more detailed than what has been summarized here; it covers every aspect of a cabling system for all applications.

A very important aspect of administration of a premise is understandability. The customer may have a particular administration system they understand, are comfortable with, and would prefer to use for their system. Knowing the customer's preference and implementing this request is more important than following the standard's recommendation. Check job specifications and converse with the customer to determine an agreed-upon labeling scheme.

ADMINISTRATION COLORS

TIA-606-C states label colors and backboards may be used to indicate the type, application, function, or position of a component within the infrastructure.

Using colors to designate the various telecommunications cables and hardware makes identifying the various subsections of a system easier. An installer looking at a wall field and seeing orange instantly knows where the service provider side of the network connection is. **See Figure 6-11.**

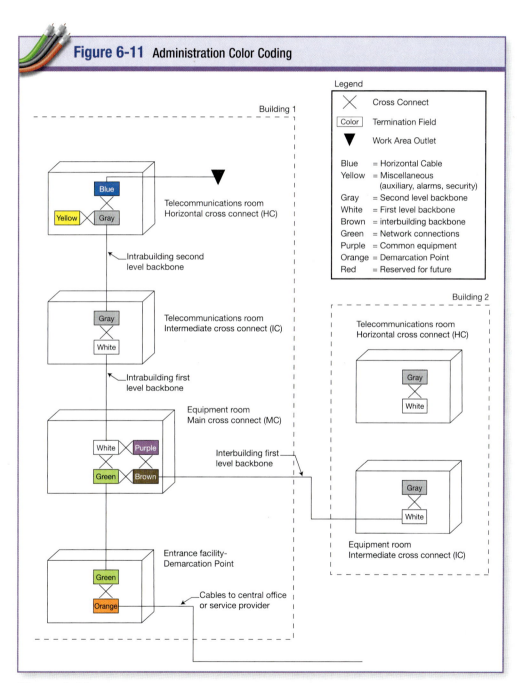

Figure 6-11 Administration Color Coding

Legend

✕	Cross Connect
Color	Termination Field
▼	Work Area Outlet

Blue = Horizontal Cable
Yellow = Miscellaneous (auxiliary, alarms, security)
Gray = Second level backbone
White = First level backbone
Brown = interbuilding backbone
Green = Network connections
Purple = Common equipment
Orange = Demarcation Point
Red = Reserved for future

Building 1

Blue
Yellow Gray
Telecommunications room
Horizontal cross connect (HC)

Intrabuilding second level backbone

Gray
White
Telecommunications room
Intermediate cross connect (IC)

Intrabuilding first level backbone

White Purple
Green Brown
Equipment room
Main cross connect (MC)

Interbuilding first level backbone

Green
Orange
Entrance facility-
Demarcation Point

Cables to central office or service provider

Building 2

Telecommunications room
Horizontal cross connect (HC)

Gray
White

Gray
White

Equipment room
Intermediate cross connect (IC)

Figure 6-11. Color codes are used to simplify administration.

Figure 6-12 Administration Color Code

Label Color	Termination Type
Color Code for Labels on Termination Hardware	
Blue	Horizontal cables - Cabling Subsystem 1
White	First-level backbone cables within a building (Cabling Subsystem 3)
Gray	Second-level backbone cables within a building (Cabling Subsystem 2)
Brown	Campus cabling
Purple	Common equipment ports (switch, multiplexer, hub, etc.)
Green	Customer side of network connection
Orange	Service provider side of network connection
Yellow	Miscellaneous
Red	Reserved for future

Figure 6-12. Color coding termination hardware labels simplifies the administration of telecommunications services.

Color Coding of Termination Hardware Labels

The color coding of termination hardware labels simplifies the administration of telecommunications services. Color coding as specified in the ANSI/TIA-606-C Standard is based on the hierarchical star configuration for backbone cabling as specified in the ANSI/TIA-568.1-D Standard. That standard allows a maximum of up to two hierarchical levels in the backbone. The first level

in the hierarchy encompasses cabling from the main cross-connect to a telecommunications room or to an intermediate cross-connect (IC).

The second level encompasses cabling between two telecommunications rooms within a building or between an intermediate cross-connect and a telecommunications room in a building.

It is important for installation and administration personnel to distinguish between first-level and second-level backbone cables. Well-organized color-coding helps to identify the level of a backbone cable and ensures that two levels of backbone cabling (the maximum allowed) are not exceeded. **See Figure 6-12.**

Termination labels identifying the two ends of the same cable shall be of the same color.

Cross-connections are generally made between termination fields of two different colors.

Figure 6-13 66-Type Colored Backboard

Purple   Blue   Purple Field   Blue Field

Figure 6-13. 66-type blocks may have blue or purple wall fields indicating the cabling subsystem or cabling use.

Figure 6-14 Orange Field

Figure 6-14. The red covers on the lower part of the block are used to identify digital circuits such as ISDN or T1 circuits.

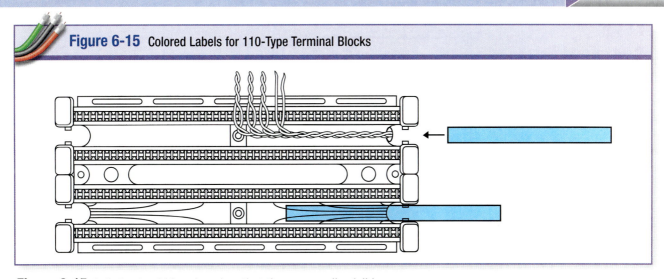

Figure 6-15 Colored Labels for 110-Type Terminal Blocks

Figure 6-15. Labels should be placed so that they are easily visible.

Examples of Administration Colors

Horizontal cables will typically have a blue backboard or blue labels to identify them. A purple field near the blue field will indicate the common equipment ports and allow for short cross-connections. **See Figure 6-13.** An orange wall field will indicate to the installer at a glance which side is the service provider's side of the demarcation point. **See Figure 6-14.**

LABELING PROCEDURES

All labels shall be printed or generated by a mechanical device. The size, color, and contrast of all labels should be selected to ensure that the identifiers are easily read. Labels should be visible during the installation and normal maintenance of the infrastructure. **See Figure 6-15.** Labels should be resistant to the environmental conditions at the time of installation (depending on the presence of moisture, heat, ultraviolet light, and so on) and should have a design life equal to or greater than that of the labeled component.

ADMINISTRATION RECORDS

Administration records are used to capture the inventory of cabling infrastructure elements and their identification.

Annexes to the TIA-606-C Standard provide many suggestions and examples regarding administration records.

Floor 1, TR A, Terminal Block D

1A-D-001	1A-D-007
1A-D-002	1A-D-008
1A-D-003	1A-D-009
1A-D-004	1A-D-010
1A-D-005	1A-D-011
1A-D-006	1A-D-012

A TIA-606-C standard-compliant 66-block label includes the TR name, block name, and termination position.

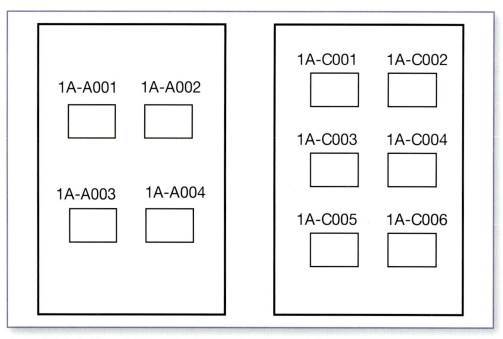

Properly labeled faceplates must adhere to the TIA-606-C Standard.

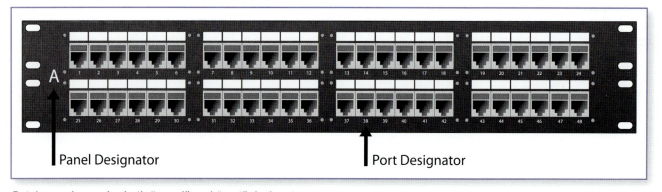

Panel Designator

Port Designator

Patch panels require both "panel" and "port" designators.

Floor 1, TR A, Terminal Block D

1A-D-001	1A-D-002	1A-D-003	1A-D-004	1A-D-005	1A-D-006
1A-D-007	1A-D-008	1A-D-009	1A-D-010	1A-D-011	1A-D-012
1A-D-013	1A-D-014	1A-D-015	1A-D-016	1A-D-017	1A-D-018
1A-D-019	1A-D-020	1A-D-021	1A-D-022	1A-D-023	1A-D-024

A TIA-606-C standard-compliant 110-block label includes the TR name, block name, and termination position.

```
WARNING
THIS OPENING HAS BEEN SEALED WITH (MANUFACTURER) FIRE STOP
PRODUCTS


IF THIS SEAL REQUIRES RETROFIT OR IS DAMAGED RESEAL WITH
(MANUFACTURER) FIRESTOP PRODUCTS ONLY


INSTALLED BY_____


UI # _____    RATING _____    DATE _____
```

Firestop labeling helps to ensure that the firestopping has been properly in-stalled or repaired.

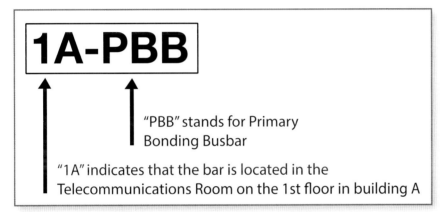

Busbar labeling should include room location and type.

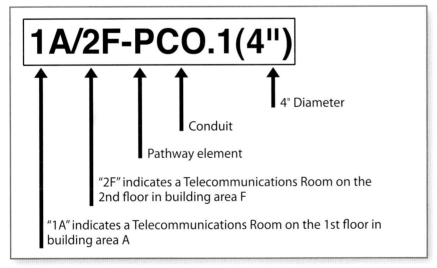

Optional pathway identifiers are placed on conduit to identify source and destination and to easily identify them among other conduits.

SUMMARY

Proper administration of a structured cabling system cannot be overstated. Administration elements vary based on the complexity of the cabling system and building infrastructure. Class 1, having only a single telecommunications space to administer, is much less involved than Class 4, with multiple building sites, campuses, and/or data centers. Using a color code system can simplify the maintenance and administration by making the cabling system more intuitive. To properly administer an infrastructure, the installer must be familiar with the TIA-606-C standard. Perhaps more importantly, however, the Electrical Worker must have good communication with the customer to determine the labeling system they prefer to use. The administration of a system does not end after the initial installation. Keeping good records of MAC work and service calls is also an important task.

REVIEW QUESTIONS

1. Which of the following is the TIA standard for labeling and documentation?
   a. 568.1-D
   b. 569-E
   c. 606-C
   d. 607-C

2. What Class of administration provides for the telecommunications needs of a single building served by multiple telecommunications spaces?
   a. Class 1
   b. Class 2
   c. Class 3
   d. Class 4

3. For the administration of firestopping systems, what is the minimum Class of administration that would be used?
   a. Class 1
   b. Class 2
   c. Class 3
   d. Class 4

4. The grounding and bonding element identifiers changed with the update to TIA-606-C. What is the new term for the telecommunications grounding busbar?
   a. BBC
   b. PBB
   c. RBB
   d. SBB

5. Which of the following is not an element to be administered in the TIA-606-C standard for a Class 1 Administration?
   a. Cabling pathways
   b. Patch panels
   c. PBB
   d. Telecommunications spaces

6. The TIA-606-C standard recommends using a grid coordinate system in a telecommunications space with multiple rows of cabinets or racks, such as computer rooms and equipment rooms, to identify the equipment cabinets and racks located within the room.
   a. True
   b. False

7. The TIA-606-C standard requires all racks and cabinets to be labeled on the __?__ and __?__ of the rack or cabinet in plain view.
   a. front / back
   b. left / right
   c. north / south
   d. top / bottom

8. Cabling Subsystem 1 links terminated on an equipment outlet are required to use the format __?__.
   a. *fs-ak*
   b. *fs.x1y1-r1*
   c. *fs=MBN*
   d. *fs-PBB*

9. Which of the following formats would be used for labeling the mesh bonding network in a grounding and bonding administration?
   a. *fs-ak*
   b. *fs.x1y1-r1*
   c. *fs=MBN*
   d. *fs-PBB*

10. Which color of label or backboard would be seen on a termination block for Cabling Subsystem 1?
    a. Blue
    b. Brown
    c. Gray
    d. White

Grounding and Bonding

Introduction

There are two aspects to consider regarding telecommunications system grounding and bonding: safety of personnel, and performance of the equipment and cabling system. For grounding of telecommunications systems for safety, look to the *National Electrical Code*. For performance grounding and bonding, look to TIA-607-C, *Generic Telecommunications Bonding and Grounding (Earthing) for Customer Premises*. However, on their own, these two documents do not ensure a quality installation. To address this issue, NECA and BICSI developed a joint standard, ANSI/NECA/BICSI-607, *Standard for Telecommunications Bonding and Grounding Planning and Installation Methods for Commercial Buildings*.

Objectives:

- Describe the five elements of a standards-based telecommunications bonding and grounding system.
- State requirements for a PBB, an SBB, a TBB, a TBC, and a BBC.
- Describe the functionality of a PBB, an SBB, a TBB, a TBC, and a BBC.
- Define *bonding conductor* and describe its purpose.
- Label grounding system components for a multi-story facility.
- State the requirements for and describe the functionality of a CBB and a CTBB, and explain when they are needed.

Chapter 7

Table of Contents

GROUNDING AND BONDING OF STRUCTURED CABLING SYSTEMS

Proper grounding and bonding of structured cabling systems is necessary for the safety of personnel and the performance of electronic equipment and cables. A safe grounding and bonding system for a building is achieved by following the requirements set forth by the *National Electrical Code (NEC)*. To achieve proper performance of electronic equipment and cables, it is essential that an installer understand the requirements of a separate grounding and bonding infrastructure system for telecommunications as set forth by TIA-607-C, *Generic Telecommunications Bonding and Grounding (Earthing) for Customer Premises.* **See Figure 7-1.**

Figure 7-1 Commercial Grounding and Bonding Infrastructure

Figure 7-1. Grounding and bonding ensures proper safety and performance. *Courtesy of Panduit Corporation*

Definitions

An installer must clearly understand several terms that are often used when grounding and bonding. Failure to have a complete understanding of the terms associated with grounding and bonding may result in improperly applying the procedures necessary to achieve a proper installation. The following definitions are from the *NEC*, Article 100, Definitions:

> **Grounded (Grounding).** Connected (connecting) to ground or to a conductive body that extends the ground connection.
> **Bonded (Bonding).** Connected to establish electrical continuity and conductivity.
> **Intersystem Bonding Termination.** A device that provides a means for connecting intersystem bonding conductors for communications systems to the grounding electrode system.

Reproduced with permission of NFPA from NFPA 70®, *National Electrical Code®* (NEC®), 2020 edition. Copyright© 2019, National Fire Protection Association. For a full copy of the NEC®, please go to www.nfpa.org.

Grounding

Grounding occurs when a connection has been established to earth. Connection to earth is established using a grounding electrode permitted by Article 250.52 of the *NEC*. The Exception to Article 250.53(A)(2) of the *NEC* requires that when the connection to earth is established using a single rod, pipe, or plate electrode, the resistance between the electrode and earth shall be 25 ohms or less. However, Article 250.50 requires that if there are multiple electrodes available at a building or structure, such as building steel, concrete-encased reinforcing bars, and metal water pipe, then they shall be bonded together to form a grounding electrode system. Such an electrode system is then assumed to provide a 25-ohm or less connection to earth. Should an installer be required to verify the resistance of an installed electrode or electrode system, a thorough understanding of earth resistance test equipment and operating procedures is needed to accurately measure the resistance to earth of an installed electrode.

Bonding

Bonding is required to reduce or eliminate hazardous potential differences between non–current-carrying metallic components within a building or structure. Components that require bonding to reduce potential differences include, but are not limited to, metallic raceways, equipment racks, enclosures, cabinets, building steel, metallic piping, and access floor structures.

A *bonding conductor* is a conductor used specifically for the purpose of bonding. Bonding conductors are not intended to carry electrical load currents under normal conditions, but they must be capable of carrying fault currents that may result from transient voltages, power faults, or lightning strikes. Bonding conductors shall be copper and insulated with a green jacket or marked with a green color using green vinyl tape. Bonding conductors should be insulated when installed in buildings to avoid unintentional ground connections. They must be installed to provide the shortest distance between points and are to be kept as straight as possible with few bends or changes in direction. If bends in a bonding conductor must be made, the minimum bend radius of eight times the bonding conductor's diameter must be maintained to ensure the bonding conductor will properly carry high levels of fault current. The *NEC* and TIA-607-C suggest 6 AWG copper as the minimum size for bonding conductors, with larger sizes sometimes required based on distance and expected fault currents.

In addition to the safety grounding and bonding system for the building or structure that is required by the *NEC* to ensure a proper level of performance, an installer must also follow the guidelines established in TIA-607-C for a separate telecommunications grounding and bonding infrastructure. TIA-607-C establishes a grounding and bonding infrastructure for small or large commercial buildings.

BASIC SYSTEM REQUIREMENTS

The TIA-607-C Standard specifies the requirements for a uniform telecommunications grounding and bonding infrastructure that shall be followed within commercial buildings where telecommunications equipment will be installed. The standard also specifies the telecommunications grounding and bonding infrastructure and its interconnection to other building systems. **See Figure 7-2.**

The basic requirements are:

- A ground (earth) reference for telecommunications systems within the telecommunications entrance facility, all telecommunications rooms, and the equipment room(s)
- Bonding of metallic pathways, cable shields, conductors, and hardware at telecommunications rooms, equipment rooms, and entrance facilities
- Proper labeling of each telecommunications grounding and bonding conductor

It is important to understand what the basic requirements of a grounding and bonding system are. However, it is equally important to understand what the TIA-607-C does not cover before discussing what the standard does cover. The TIA-607-C does not cover:

- Specific grounding and bonding of any telecommunications equipment and associated wiring
- Values of surge current immunity and insulation withstand voltages
- Methods for verifying and maintaining grounding and bonding networks
- Specific methods for RFI/EMI mitigation for equipment or systems
- Primary protector/arrester specifications, applications, and installation
- Secondary protector specifications and applications
- Specific user safety
- Grounding and bonding practices of the local telecommunications utilities
- Electrical service entrance

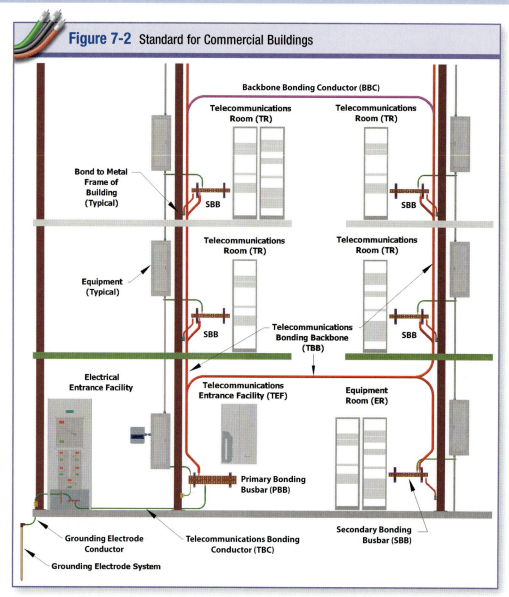

Figure 7-2 Standard for Commercial Buildings

Backbone Bonding Conductor (BBC)

Telecommunications Room (TR)

Telecommunications Room (TR)

Bond to Metal Frame of Building (Typical)

SBB

Telecommunications Room (TR)

Telecommunications Room (TR)

Equipment (Typical)

SBB

Telecommunications Bonding Backbone (TBB)

SBB

Electrical Entrance Facility

Telecommunications Entrance Facility (TEF)

Equipment Room (ER)

Primary Bonding Busbar (PBB)

Secondary Bonding Busbar (SBB)

Grounding Electrode Conductor

Telecommunications Bonding Conductor (TBC)

Grounding Electrode System

Figure 7-2. *The scope of TIA-607-C states that the standard specifies requirements for a generic telecommunications grounding infrastructure and its interconnection to electrical systems and telecommunications systems.* Courtesy of Harger Lightning and Grounding

- Grounding of AC surge protection devices
- Buildings with more than one electrical service entrance

THE SYSTEM

The basic components of a grounding and bonding system as specified in TIA-607-C include:

1. Primary bonding busbar (PBB)
2. Telecommunications bonding conductor (TBC)

Larger installations may also include:

3. Telecommunications bonding backbone (TBB)
4. Secondary bonding busbar (SBB)
5. Backbone bonding conductor (BBC)

Primary Bonding Busbar (PBB)

The primary bonding busbar (PBB) serves as the dedicated extension of the building grounding electrode system for the telecommunications infrastructure. (Formerly, the PBB was called the *telecommunications main grounding busbar.*) The PBB also serves as the central attachment point for the telecommunications bonding backbone(s) (TBB) and equipment. There is a single PBB per building located in the telecommunications entrance facility.

The PBB shall be mounted with an insulated support that separates the PBB from the wall by 50 millimeters (2 in). The PBB shall be six millimeters (0.25 in) thick by 100 millimeters (4 in) wide, and is variable in length based on connections needed. The PBB shall be listed and made of copper or copper alloys with a minimum of 95% conductivity.

All connections of conductors to the PBB shall be made using exothermic welding or irreversible compression type connections. When using irreversible compression type connections, listed two-hole lugs with correctly-matched connecting hardware shall be used.

All metallic raceways for the telecommunications cabling located within the same room or space as the PBB shall be bonded to the PBB. Where the metal frame (structural steel) of the building is readily accessible within the room or space, the PBB shall be bonded to the steel frame using a minimum 6 AWG conductor. When an electrical panelboard is located in the same room as the PBB, the panelboard's alternating current equipment ground (ACEG) bus or the enclosure shall be bonded to the PBB. **See Figure 7-3.**

Telecommunications Bonding Conductor (TBC)

The telecommunications bonding conductor (TBC) shall bond the PBB to the service equipment (power) ground. (The TBC was formerly known as the *bonding conductor for telecommunications*.) The telecommunications bonding conductor shall be at minimum the same size as the TBB. This connection may be completed by the electrical contractor. **See Figure 7-4.**

Telecommunications Bonding Backbone (TBB)

The telecommunications bonding backbone (TBB) is a conductor that interconnects all secondary bonding busbars (SBBs) with the PBB. The TBB originates

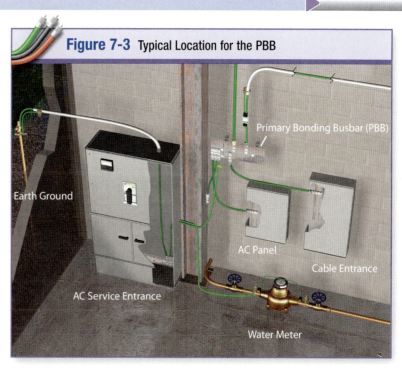

Figure 7-3 Typical Location for the PBB

Figure 7-3. *The primary bonding busbar (PBB) is usually bonded to the steel frame of the building, if accessible.* Courtesy of Panduit Corporation

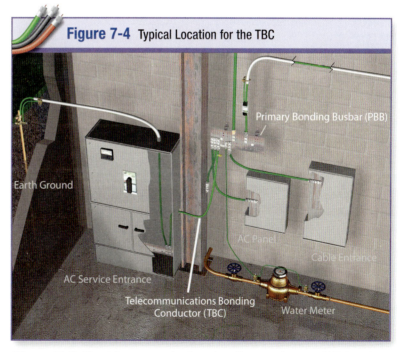

Figure 7-4 Typical Location for the TBC

Figure 7-4. *The telecommunications bonding conductor (TBC) bonds the PBB to the electrical service equipment.* Courtesy of Panduit Corporation

at the PBB and extends throughout the building, using the telecommunications backbone pathways to connect all of the SBBs in the telecommunications rooms and equipment rooms. **See Figure 7-5.**

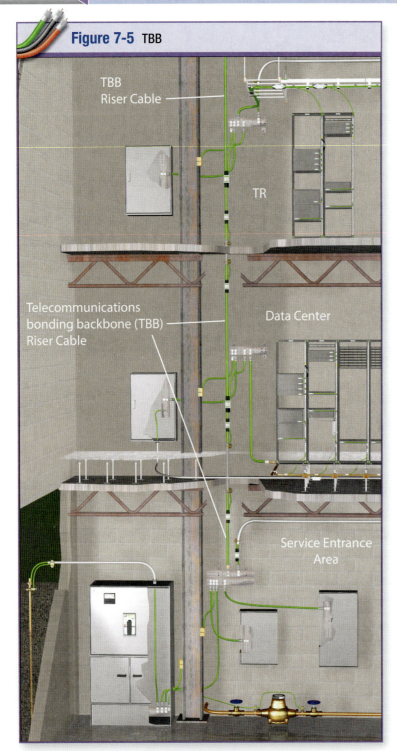

Figure 7-5 TBB

TBB Riser Cable

TR

Telecommunications bonding backbone (TBB) Riser Cable

Data Center

Service Entrance Area

Figure 7-5. *The telecommunications bonding backbone (TBB) connects all secondary bonding busbars (SBBs) in the building with the primary bonding busbar (PBB).* Courtesy of Panduit Corporation

Figure 7-6 Sizing the TBB/BBC

TBB/BCC linear length m (ft)	Conductor size (AWG)
Less than 4 (13)	6
4 – 6 (14 – 20)	4
6 – 8 (12 – 26)	3
8 – 10 (27 – 33)	2
10 – 13 (34 – 41)	1
13 – 16 (42 – 52)	1/0
16 – 20 (53 – 66)	2/0
20 – 26 (67 – 84)	3/0
26 – 32 (85 – 105)	4/0
32 – 38 (106 – 125)	250 kcmil
38 – 46 (126 – 150)	300 kcmil
46 – 53 (151 – 175)	350 kcmil
53 – 76 (176 – 250)	500 kcmil
76 – 91 (251 – 300)	600 kcmil
Greater than 91 (301)	750 kcmil

Figure 7-6. *Use this chart for sizing of the TBB or the BBC based on length.*

insulated TBB shall have its insulation meet the fire ratings of its pathway. Sizing of the TBB should be based on Table 1 in TIA-607-C. **See Figure 7-6.**

The TBB should be installed without splices. Where splices are necessary, the number of splices should be kept at a minimum, and the splices shall be accessible in a telecommunications space. The TBB and splices shall be protected from damage.

Secondary Bonding Busbar (SBB)

The secondary bonding busbar (SBB) is the grounding connection point for the telecommunications systems and equipment in the area served by the telecommunications room or equipment room. **See Figure 7-7.** (The SBB was formerly known as the *telecommunications grounding busbar*.)

The SBB shall be mounted with an insulated, 50-millimeter (2-in) support that separates the SBB from the wall. The SBB shall be six millimeters (0.25 in) thick by 50 millimeters (2 in) wide and is variable

The TBB shall be a copper conductor and minimally sized as a 6 AWG conductor with a maximum size of 750 kcmil, depending on the length cabling the TBB. An

Figure 7-7 SBB Location

Figure 7-7. *Each telecommunications room has a secondary bonding busbar (SBB), which serves to ground all telecommunications systems and equipment in the area.* Courtesy of Panduit Corporation

in length. All connections of the conductors to the SBB shall be made using exothermic welding or irreversible compression type connections. When using irreversible compression type connections, listed two-hole lugs with correctly-matched connecting hardware shall be used.

All metallic raceways for the telecommunications cabling located within the same room or space as the SBB shall be bonded to the SBB. Where the metal frame (structural steel) of the building is readily accessible within the room or space, the steel frame shall be bonded to the SBB using a minimum 6 AWG conductor. When an electrical panelboard is located in the same room as the SBB, the panelboard's ACEG bus or the enclosure shall be bonded to the SBB.

Backbone Bonding Conductor (BBC)

The backbone bonding conductor (BBC) is the conductor that interconnects elements of the telecommunications grounding infrastructure (formerly the *grounding equalizer*). The BBC is utilized in larger, multi-story commercial buildings with multiple telecommunications backbone risers. Multiple TBB risers may develop potential differences between one another as they route through the building. Such potential differences may

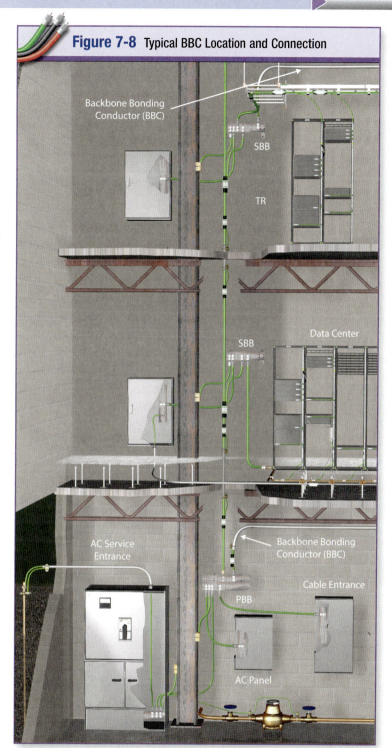

Figure 7-8 Typical BBC Location and Connection

Figure 7-8. *The backbone bonding conductor (BBC) equalizes differences caused by varying levels of EMI throughout the building.* Courtesy of Panduit Corporation

be the result of varying levels of EMI within the building and the difference in length. The BBC is used to equalize those potential differences. **See Figure 7-8.** When there are two or more TBB risers, the TBBs shall be bonded together with a

Figure 7-9 Proper Labeling at the SBB

Figure 7-9. Labeling standards ensure that every part of the grounding system is in place and accounted for. Courtesy of Panduit Corporation

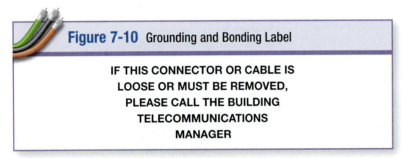

Figure 7-10 Grounding and Bonding Label

IF THIS CONNECTOR OR CABLE IS
LOOSE OR MUST BE REMOVED,
PLEASE CALL THE BUILDING
TELECOMMUNICATIONS
MANAGER

Figure 7-10. Labels for grounding and bonding systems must include the above information.

BBC at the top floor and at a minimum of every third floor in between to the lowest level.

Within a large single-story building where more than one TBB is used, they shall be connected with a BBC at the location farthest from the PBB and every 10 meters (33 ft) back to the PBB. The sizing of the BBC is to be determined using the same table that is to be used for sizing the TBB. Connections of the BBC to the TBB may be made at the SBB utilizing exothermic welding or irreversible compression type connectors.

The telecommunications bonding conductor (TBC), each telecommunications bonding backbone (TBB) conductor, and each backbone bonding conductor (BBC) shall be green, be marked with a distinctive green color, or be green with a yellow stripe.

Labeling

The TIA-607-C Standard requires each bonding conductor to be labeled. **See Figure 7-9.** Both the TIA-607-C and the ANSI/TIA-606-C (Administration) standards provide guidelines for labeling. Each telecommunications grounding and bond-

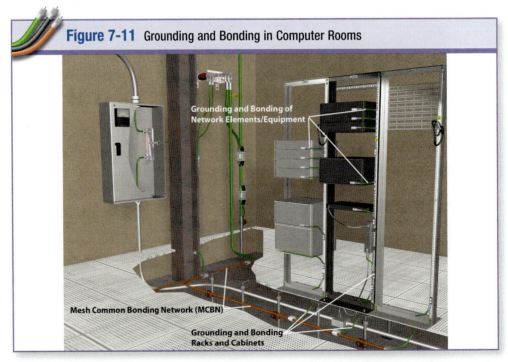

Figure 7-11 Grounding and Bonding in Computer Rooms

Grounding and Bonding of Network Elements/Equipment

Mesh Common Bonding Network (MCBN)

Grounding and Bonding Racks and Cabinets

Figure 7-11. Small data centers must have a common bonding network, typically a mesh bonding network, that connects all conductive surfaces to each other and to the Earth. Courtesy of Panduit Corporation

ing conductor shall be labeled, and the label shall be located in a readable position as close as practicable to the conductor's point of termination. Labels shall be nonmetallic and contain required information. **See Figure 7-10**. Refer to ANSI/TIA-606-C for additional labeling requirements.

Computer Rooms

Each computer room shall contain an SBB (or PBB if so designed) and should also contain a supplementary bonding network that is bonded to the SBB or PBB. **See Figure 7-11.**

The supplementary network is typically a mesh bonding network (mesh-BN). The mesh-BN provides a greater degree of equipotential bonding by connecting all conductive surfaces of a computer room to each other and to the Earth. This process is essential to prevent stray voltage that could affect vital equipment and potentially harm personnel.

The supplementary bonding network could also be a mesh isolating bonding network (IBN) that is isolated from the common bonding network except for at one controlled point, or a star IBN that is deployed into a star instead of a mesh.

Each cabinet and rack within the computer room shall have its own, dedicated bonding conductor to connect it to the PBB, SBB, telecommunications equipment bonding conductor, or mesh-BN with a minimum sized 6 AWG conductor. **See Figure 7-12.**

Network Equipment Grounding and Bonding

Per the ANSI/TIA-607-C Standard, grounding the telecommunications equipment through the AC power cord (even though it is typically equipped with a grounding or bonding wire) does not meet the intent of the standard. It is intended that the information technology equipment be provided a specific, supplementary ground path in addition to the required AC ground path. Many types of equipment do not require additional bonding conductors and do not have attachment points for them; this equipment may be bonded either through the equipment rail or through the power cord.

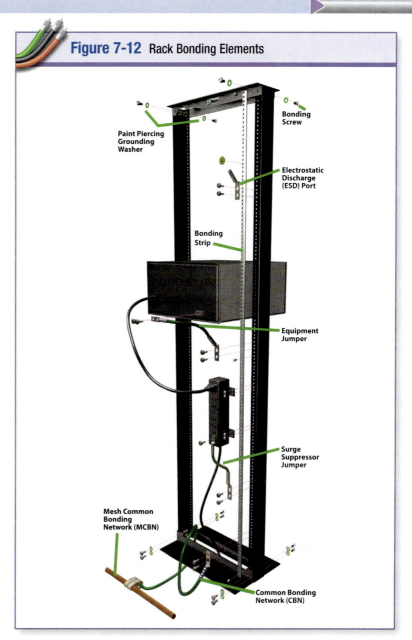

Figure 7-12 Rack Bonding Elements

Paint Piercing Grounding Washer

Bonding Screw

Electrostatic Discharge (ESD) Port

Bonding Strip

Equipment Jumper

Surge Suppressor Jumper

Mesh Common Bonding Network (MCBN)

Common Bonding Network (CBN)

Figure 7-12. Requirements for rack bonding elements can be found in TIA-607-C.

Telecommunications Equipment Bonding Conductor (TEBC)

The telecommunications equipment bonding conductor (TEBC) connects the equipment racks/cabinets to the PBB or SBB within the telecommunications space. There will be one continuous copper conductor, sized not less than 6 AWG, for each rack or cabinet.

The TEBC will be connected to the rack or cabinet via a vertical or horizontal rack bonding busbar (RBB) using listed, irreversible two-hole lugs or listed

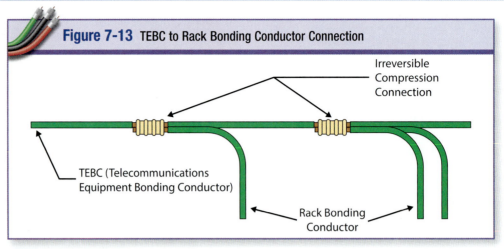

Figure 7-13 TEBC to Rack Bonding Conductor Connection

Irreversible
Compression
Connection

TEBC (Telecommunications
Equipment Bonding Conductor)

Rack Bonding
Conductor

Figure 7-13. *The TEBC to rack bonding conductor connection must be made with an irreversible compression type connector.*

terminal blocks with two internal hex screws. Connections to the TEBC shall be made with listed irreversible compression type connectors suitable for multiple conductors. When making rack bonding conductor (RBC) connections to the TEBC, the RBC connections shall be routed toward the PBB or SBB. **See Figure 7-13.** The TEBCs may be routed in or on the side of cable trays or ladder racks or under access floors. They shall be supported at intervals of no more than 0.9 meters (3 ft) and shall be separated from power or telecommunications cables by 50 millimeters (2 in).

Rack Bonding Busbar (RBB)

The rack bonding busbar shall be listed and have the cross-sectional area equivalent to or exceeding that of a 6 AWG wire. The RBB shall be bonded to the rack and to the TEBC. Connections to the RBB shall be made with exothermic welding or listed irreversible compression two-hole lugs. The telecommunications equipment connected to the RBB shall be connected using listed compression connectors to the grounding post of the equipment when provided.

MULTI-TENANT BUILDINGS

TIA-607-C-1, *Generic Telecommunications Bonding and Grounding (Earthing) for Customer Premises, Addendum 1 – Bonding in Multi-tenant Buildings (January 2017)* adds instructions for shared services between tenants to TIA-607-C.

With today's smart buildings, a common bonding infrastructure provides the ability to equalize potential differences between shared services such as building automation systems and distributed antenna systems. It also may provide a bonding infrastructure to which separate tenants are connected. If a common bonding infrastructure is present, it should include a common bonding busbar (CBB), a telecommunications bonding conductor (TBC), and also a common telecommunications bonding backbone (CTBB).

The CBB is a dedicated extension of the grounding electrode system for telecommunications and provides an attachment place for the CTBBs and the equipment. In a multi-tenant building, the CBB serves the same purpose as a PBB would in a single-tenant building. The CBB must meet the same requirements as stated in the ANSI/TIA-607-C standard for the PBB. If there is no common bonding infrastructure, each tenant's PBB shall be bonded to the main electrical panel.

The telecommunications bonding conductor (TBC) connects the common bonding busbar to the electrical service (power) ground. In a multi-tenant building, this conductor has the same purpose as it has in a single-tenant building.

The common telecommunications bonding backbone (CTBB) is a conductor that connects all of the individual tenants' PBBs to the CBB and also connects all SBBs (used as a bonding point for shared telecommunications services and equipment in an area served by a common distributor room) with the CBB. This is intended to equalize all potential differences between the separate tenants and the separate building services. The CTBB shall meet the requirements for a TBB in the ANSI/TIA-607-C standard. **See Figure 7-14.**

Figure 7-14 Example of a Multi-Tenant Building

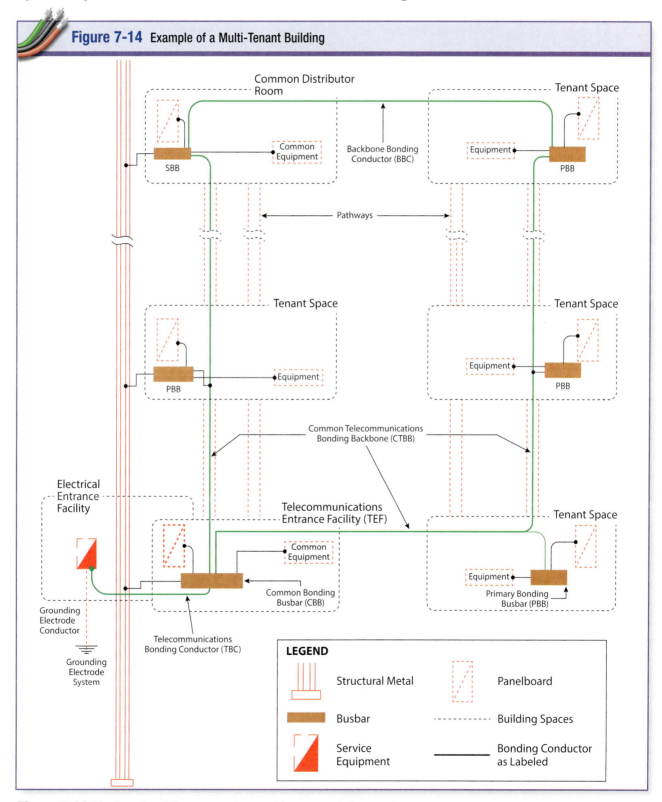

Figure 7-14. *The bonding infrastructure in a multi-tenant building provides equalization of potential for shared services.*

STP COPPER SYSTEMS

Grounding and bonding is the single most important practice for the protection of people and property. Throughout the industry, there is a misconception that grounding and bonding of shielded twisted-pair (STP) copper cabling systems is more difficult, more time consuming, and more technique sensitive, but this is simply not true. Properly grounding and bonding an STP copper cabling system is straightforward and essentially no different than grounding and bonding an unshielded twisted-pair (UTP) system. The grounding and bonding requirements specified in TIA-607-C, *Generic Telecommunications Bonding and Grounding (Earthing) for Customer Premises,* are applicable to both STP and UTP systems.

STP cable and connectivity have been on the market for many years and have been widely deployed and proven throughout the world. STP has always offered better RFI/EMI immunity, higher security, and an electrically superior performance. With 10 Gigabit Ethernet over copper now a reality, the performance of STP systems offers significantly more headroom over 10 Gigabit UTP systems. Furthermore, STP offers a smaller cable diameter and better fill ratio, providing higher density and cost savings.

STP in the TR

In each TR, any metallic component that is attached to the telecommunications cabling infrastructure, or comes in contact with any other metallic component attached to the infrastructure, must be bonded to the secondary bonding busbar (SBB) per the TIA standard. That includes racks, ladders, enclosures, metallic cable trays, and equipment like patch panels, routers, and switches. This process is exactly the same for both STP and UTP systems.

Shielded patch panels, jacks, and cable have the grounding and bonding built in. So once the cable is properly terminated to the jack and the jack mounted to the patch panel, bonding and grounding is complete.

STP at the Workstation

Grounding and bonding STP at the workstation is also relatively straightforward. In a permanent link, the shield is bonded only at the TR end as previously described. The workstation end of the permanent link is not bonded to ground. This is the same for both UTP and STP systems.

In a channel, the bonding of the permanent link shield to ground at the workstation end is accomplished by connecting a shielded patch cord between the outlet and the equipment. This results in a ground condition at both ends of the channel. The equipment is grounded through the electrical receptacle, tying both the electrical and telecommunications system to ground. **See Figure 7-15.**

Maintaining Continuity of the Shield

In an STP system, the shield should completely surround the cable along its entire length, and the shield should remain continuous along the entire length of the channel. This is accomplished by using only shielded products throughout the entire channel, from cable to plug to jack to patch panel.

Proper installation is also central to maintaining continuity of the shield. When terminating a shielded jack, the shield must make contact with the connector. Today's shielded connectivity hardware is made completely out of metal, ensuring that contact is quite simple. For example, some shielded jacks feature a spring-loaded cable strain relief that automatically provides a 360° contact to the terminated shield.

Proper grounding and bonding is a sizeable job whether it is UTP or STP, and there is virtually no difference in the requirements between the two. The shield must maintain continuity throughout a channel, but achieving that condition is as simple as properly terminating the shielded jack and using shielded components throughout the channel. In today's IT world, it is vital for customers who rely heavily on their communication systems

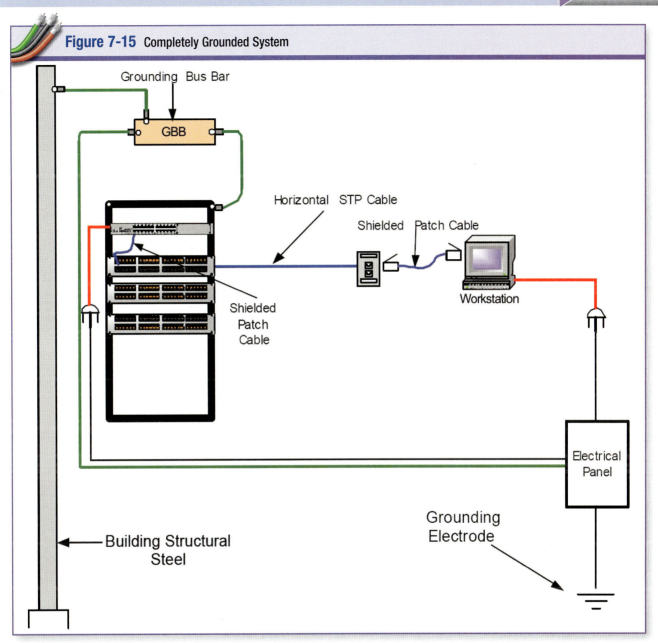

Figure 7-15 Completely Grounded System

Figure 7-15. Maintaining the grounding connection throughout the system requires the use of STP, shielded jacks, and shielded patch cables. Courtesy of Tyco Electronics™

to ensure proper grounding and bonding – whether it is a UTP or STP system.

TESTING FOR GROUND LOOP AND PROPER GROUNDING AND BONDING

In an electrical system, a *ground loop* refers to an unwanted current that flows in a conductor and connects two points that are at different voltages. Another misconception surrounding shielded cable is that ground loops will always be a problem. However, ground loops that affect network performance can only occur when a system has more than one path to ground and the voltage difference between the two points is more than one volt. If the components of a telecommunications system are properly bonded and the system is effectively bonded to the building's grounding electrode system, there is virtually no condition where a ground loop is possible.

The only way to ensure proper grounding and bonding for either a UTP or STP system is to test them upon completion. Ground potential difference should not exceed one volt. Testing for ground potential difference can be done using an earth ground resistance tester with the entire building in operation. In other words, nothing needs to be shut down to test the grounding and bonding. Earth ground resistance testers can be purchased almost anywhere telecommunications testers are sold.

It is imperative to test the potential difference between the PBB and the electrical ground and between the PBB and each SBB. Within the telecommunications space, testing should also be done between the SBB and all racks, cable tray,

and electronic equipment. TIA-568.0-D 6.2.5 states that voltage greater than one volt rms between the screen of Cabling Subsystem 1 at the equipment outlet (EO) and the ground wire of the corresponding electrical outlet expected to provide power to the equipment indicates improper grounding and is not recommended. The bonding and grounding infrastructure should be modified so that the voltage is less than one volt rms. Simply put, the difference in ground potential between the telecommunications and electrical system at the workstation should not exceed one volt.

Some may want to have their grounding and bonding system independently tested or certified by a qualified electrical contractor or engineer. Locating an independent grounding and bonding test engineer may be difficult, but if a certified test is requested, only use qualified consultants.

SUMMARY

The installer must be aware of the reasons for grounding (connecting to ground or to a conductive body that extends the ground connection) and bonding (connecting to establish electrical continuity and conductivity) telecommunications systems and the relationships of grounding and bonding to safety and performance. It is often argued that these low-voltage (50 volts or less) and/or no voltage (optical fiber) systems pose no safety hazard, and therefore the *NEC* need not be so restrictive on these systems. However, telecommunications system power is most often supplied by the building AC supply. This means that at some point, a telecommunications device or system is connected to building power and, by association, the building's grounding system. Telecommunications systems are not immune from carrying a fault current or a surge from a lightning strike, so from a safety standpoint, all systems connected to a grounding system must meet the same safety requirements.

Performance issues usually arise from improper grounding and bonding that leads to stray voltages and improper signal reference grounds. Between the power system and the telecommunications system, there are lots of parallel paths to create a difference in potential and for stray currents to flow. High resistance connections in the grounding or bonding system can help to create voltage drops that would exceed the maximum ground potential difference at the work station of one volt. For example, a small one-milliampere current flowing through a poor ground connection at a work station of 120 ohms would cause a voltage drop of 1.2 volts across the connection ($E = I \times R$ or $E = I \times Z$). Although only 0.2 volts above the maximum, this is still enough to possibly cause bit errors in the signal. Signal quality is a direct function of good grounding and bonding techniques.

REVIEW QUESTIONS

1. Grounding requirements for structured cabling systems are only covered in Chapter 8 of the *NEC*.
   a. True
   b. False

2. The *NEC* requires that when the connection with earth is established using a single ground rod, pipe, or plate electrode, the resistance between the electrode and earth shall be __?__ or less.
   a. 15 Ω
   b. 25 Ω
   c. 50 Ω
   d. 100 Ω

3. Which of the following is not covered by TIA-607?
   a. Values of surge current immunity and insulation withstand voltages
   b. Primary protector/arrester specifications, applications, and installation
   c. Grounding and bonding practices of the local telecommunications utilities
   d. All of the above

4. The __?__ serves as the dedicated extension of the building grounding electrode system for the telecommunications infrastructure.
   a. BBC
   b. PBB
   c. TBB
   d. TBC

5. The __?__ shall bond the PBB to the service equipment ground.
   a. BBC
   b. SBC
   c. TBB
   d. TBC

6. The __?__ is a conductor that interconnects all SBBs with the PBB.
   a. BBC
   b. PSB
   c. TBB
   d. TBC

7. The __?__ is the grounding connection point for the telecommunications systems and equipment in the area served by that telecommunications room or equipment room.
   a. BBC
   b. CBB
   c. SBB
   d. TBC

8. The __?__ is the conductor that interconnects elements of the telecommunications grounding infrastructure.
   a. BBC
   b. PBB
   c. TBB
   d. TBC

9. The TEBC connects the equipment racks/cabinets to the __?__ within the telecommunications space.
   a. BBC
   b. PBB
   c. SBB
   d. either b. or c.

10. The TEBC shall be supported at intervals of no more than __?__.
    a. 0.6 m (2 ft)
    b. 0.9 m (3 ft)
    c. 1.5 m (5 ft)
    d. 3 m (10 ft)

Configuring Structured Cabling Systems

Introduction

There are a number of items that must be considered when configuring a structured cabling system. First and foremost, the type of outlet needs to be determined as well as how it will be terminated (T568A, T568B, USOC). It must be known what category UTP will be used, or if fiber will be used. Next, the amount of cable will need to be determined to serve each work area.

After the cable is pulled to the work areas and jacks are terminated, the technician will move on to the telecommunications rooms (TRs). At this point, the wall fields will be configured. Wall field configuration utilizes color coding for horizontal cables between outlets and the TR, and for first-level backbone cables, second-level backbone cables, and common equipment. Arrangement of the wall fields must also be determined at this point.

Objectives

- Determine the amount of cable needed for a structured cabling system.
- Establish a color code system for the wall field in the telecommunications room.
- Determine the type(s) of 110-type terminal blocks needed for the structured cabling system.
- Explain how to arrange a typical wall field in the telecommunications room.

Chapter 8

Table of Contents

CONFIGURING FACEPLATES

In order to configure work area faceplates, the installer must know the following:

- What categories of UTP outlets will be required?
- What pin/pair assignments (T568A or T568B) will be used?
- What color should each specific outlet be?
- How many of each color/category of outlet will be required?
- What types of fiber adapters will be required?
- How many fiber adapters will be required?
- Will STP outlets be required; if so, how many?
- Will coaxial couplings be required; if so, how many and what type?
- How many faceplates will be flush-mounted?
- How many faceplates will be surface-mounted?
- How many faceplates will be furniture-mounted? What type(s) will be required?
- What color should the faceplates be?

Once this data is known, the installer can select the appropriate quantities and types of faceplates, outlets, and adapters. To properly configure work area faceplates, it is necessary to become familiar with the guidelines contained in the ANSI/TIA-568 Standards, which pertain to telecommunications outlets (TOs). Specifically, these Standards require that all four pairs in a horizontal cable be terminated on only one outlet. Individual pairs cannot be split among more than one outlet, nor can the pairs be simultaneously terminated (daisy-chained) on more than one outlet.

Note that some networks or services require application-specific electrical components (such as impedance matching devices) on the telecommunications outlet. In some cases it is desirable to use cabling adapters to break out individual pairs from a 4-pair TO to more than one modular jack. To be compliant with the ANSI/TIA-568 Standards, these application-specific components and cabling adapters shall not be installed as part of the horizontal cabling. When needed, such electrical components shall be placed external to the telecommunications outlet in the work area.

CONFIGURING HORIZONTAL CABLING

Five steps must be followed in order to determine how much of each type of horizontal cable is required:

1. Establish areas to be served by each telecommunications room.
2. Determine cable distribution methods and cabling routes.
3. Determine types and quantities of cables to serve each work area.
4. Determine average cable length. **See Figure 8-1.**
   - Identify the shortest cable run: the TO location closest to the serving room (A).
   - Identify the longest cable run: the TO location farthest from the serving room (B).
   - Measure (A) and (B) cable runs along their cable pathways.

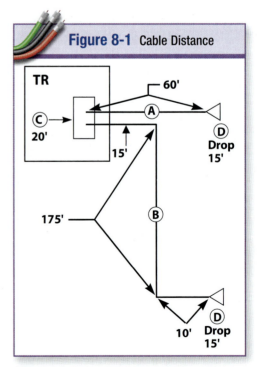

Figure 8-1 Cable Distance

Figure 8-1. Determining average cable length requires minor calculations.

Figure 8-2 Example of Cable Pull Calculation

(A)	(B)	(AL)	(S)	(C)	(D)	(TAL)
Shortest Cable Run	Longest Cable Run	Average Cable Length	10% Slack Length	TR Termination Allowance	Work Area Drop Allowance	Total Average Cable Length Termination
60 Ft.	200 Ft.	130 Ft.	13 Ft.	20 Ft. *	15 Ft. **	178 Ft.

* Variable
** Only required with overhead distribution

Figure 8-2. Average cable length helps determine the approximate amount of cable needed for the job.

- Measure the longest and shortest runs following their probable routes.
- Calculate the average cable length. (AL) = A + B ÷ 2
- Calculate slack. (S) = AL × 10% (Slack is only calculated on the (AL))
- Determine TR termination allowance (C). (Variable depending on TR size and pathway)
- Determine work area drop length (D) (Only required with overhead distribution)
- Calculate the total average cable length. (TAL) = AL + S + C + D

5. Multiply the average cable length by the number of work outlets to determine the total quantity of horizontal cable to order for each cable type (category 6, optical fiber, coaxial cable, etc.). **See Figure 8-2.**

Note: Cable runs of extreme length will skew the average length calculation if not removed from the calculation given above.

4-pair UTP cable is normally purchased in 1,000-foot lengths in boxes or on reels. If lengths greater than 1,000 feet are desired, lengths up to about 16,000 feet are available on reels. Give careful consideration to the packaging before ordering.

Pay particular attention to the average length of each wire run and the number of wire runs that can be taken from the box or reel ordered. Other cable types may be available in boxes, but they are always available on reels.

Sample Calculation for 1,000-Foot Boxes

Maximum orderable length ÷ total average length of cable run = number of runs per 1,000 ft. box.

1,000 ft. ÷ 178 ft. = 5.6 cable runs per 1,000 ft. box (rounded down to 5; partial runs are not counted)

Since each TO represents a separate cable, the total number of cable runs is equal to the total number of TOs.

Therefore, assuming there are 142 TOs to be served:

Number of TOs ÷ number of runs per 1,000 ft. box = number of boxes of cable

142 ÷ 5 = 28.4 boxes
Round up to 29 boxes

CONFIGURING A WALL FIELD IN A TELECOMMUNICATIONS ROOM

The various cabling system elements appearing on a wall field in a TR may be identified by color-coded labels.

By thinking in terms of colors, the installer can easily identify and size the terminal blocks required in the wall field. The following list identifies the termination hardware label colors that typically appear in a TR. **See Figure 8-3.**

- Blue   Horizontal cables between TOs and a TR
- White   First-level backbone cables between the main cross-connect (MC) and a TR or between the MC and an intermediate cross-connect (IC)
- Gray   Second-level backbone (tie) cables between TRs or between an IC and a TR
- Purple   Common equipment ports (switch, multiplexer, hub, etc.)

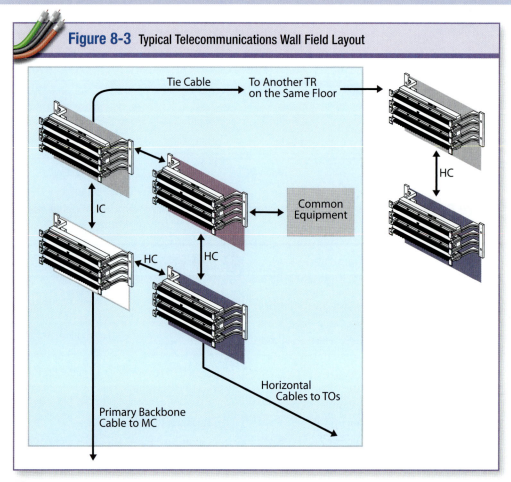

Figure 8-3 Typical Telecommunications Wall Field Layout

Tie Cable — To Another TR on the Same Floor

HC

IC

Common Equipment

HC

HC

Horizontal Cables to TOs

Primary Backbone Cable to MC

Figure 8-3. *A wall field utilizes color schemes to identify cabling and cross-connects.*

The wall field in a TR will always have blue and white fields. The gray field will only be present when tie (second-level backbone) cables are terminated in the TR. The purple field will only be present when circuits representing common equipment located within the TR are present.

Cross-connects may be made between any of the fields present in the TR. The primary purpose of these cross-connects is to assign TOs to pairs in backbone cables or to ports on common equipment. **See Figure 8-4.**

Figure 8-4 Cross-Connect Field Colors

Cross-connect	Function
Blue-to-White	Assigns a TO to an ER common equipment circuit that has been extended to the TR through backbone cabling
Blue-to-Gray	Assigns a TO served by the TR to an ER common equipment circuit that has been extended to the TR through backbone cabling via an IC
Blue-to-Purple	Assigns a TO to a common equipment circuit present in the TR
Gray-to-Purple	Assigns a TO in another TR to a common equipment circuit present in this TR
Gray-to-White	Cross-connects assigned pairs from a first-level backbone cable to assigned pairs in a second-level backbone cable

Figure 8-4. *Color codes are used to identify cross-connects on the wall field.*

Figure 8-5 110-Type Terminal Block Layout

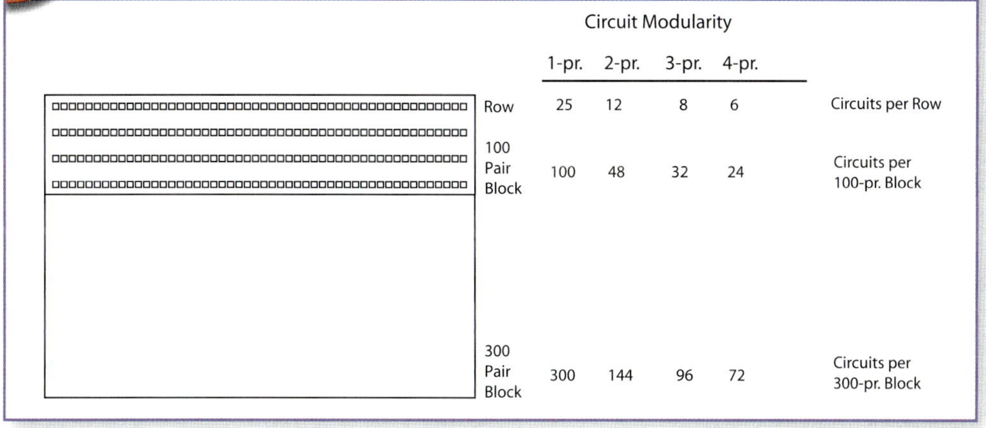

	Circuit Modularity				
	1-pr.	2-pr.	3-pr.	4-pr.	
Row	25	12	8	6	Circuits per Row
100 Pair Block	100	48	32	24	Circuits per 100-pr. Block
300 Pair Block	300	144	96	72	Circuits per 300-pr. Block

Figure 8-5. *A 100-pair 110-type terminal block can accommodate twenty-four 4-pair cables.*

SIZING 110-TYPE WALL FIELDS

Typically, 110-type terminal blocks are available in 25-, 50-, 100-, and 300-pair sizes. This type of terminal block terminates 25 pairs on each row. **See Figure 8-5.** They use 3-, 4-, or 5-pair connecting blocks depending on the number of pairs required for each circuit. A 3-pair circuit (3-pair modularity) requires 3-pair connecting blocks; a 4-pair circuit requires 4-pair connecting blocks. A 2-pair circuit can also use 4-pair connecting blocks since four is a multiple of two. The 5-pair connecting block is used when terminating multi-pair cables.

The following design guidelines are used to determine the number of terminal blocks required for each wall field color present in the TR. In order to lay out wall fields using other types of blocks (66, Krone, BIX, etc.), follow the same procedures. Only the dimensions and number of pairs per block will be different.

Backboards

The different field colors of 110-type terminal blocks are typically separated using a backboard. A backboard is a panel (for example, wood or metal) used for mounting connecting hardware and equipment. Backboards are used to hold and support cross-connect wires routed between terminal blocks. **See Figure 8-6.**

The backboard may or may not have a standoff, which allows cables to be routed behind it. Backboards should be located in the center of a wall field's height to minimize the length of cross-connect wires.

Field Colors

Section 10 in the ANSI/TIA-606 standard identifies color recommendations for a wall field. Though the color codes provided in the document are not a requirement, the standard does require consistency when alternative color codes are used.

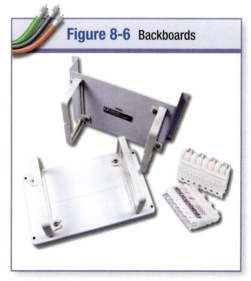

Figure 8-6 Backboards

Figure 8-6. *Typically, 188-type backboards separate 110-type terminal blocks on a wall field.*

Figure 8-7 110-Type Terminal Block Capacities

110 Terminal Block Capacity	No. of 25-Pair Rows	No. of TOs Served per Row	No. of TOs Served per Block/Assembly
100-pair Block	4	6	24
300-pair Block	12	6	72

Figure 8-7. *Each row of a 110-type block can serve six telecommunications outlets.*

Blue - Horizontal Cabling

When 4-pair horizontal cables are to be terminated on a 110-type terminal block, each cable represents a TO at the work area. Six 4-pair cables representing six TOs can be terminated on one row of a 110-type block. A 300-pair block contains twelve 25-pair rows, each row serving six TOs for a total of 72 TOs served per block assembly. **See Figure 8-7.**

The number of 110-type terminal blocks required for the blue field is based upon the total number of TOs served by the TR.

The following formulas can be used to determine the necessary quantity of 110-type terminal blocks when 4-pair cables are specified.

Number of TOs ÷ 24 = Number of 100-pair Terminal Blocks

Number of TOs ÷ 72 = Number of 300-pair Terminal Blocks

Sample Calculation

Example:
How many 300-pair 110-type terminal blocks are required to terminate 4-pair cables from 370 TOs?

Solution:

$$370 \div 72 = 5.14$$

Therefore, six 300-pair 110-type terminal blocks are required.

Note that these calculations must be performed by dividing the number of TOs by 24 or 72. Multiplying the number of TOs by four and dividing by 100 or 300 will yield inaccurate results because the 25th pair position of each row is not used when terminating 4-pair cables.

In the example given, multiplying 370 times four (which equals 1,480) and dividing it by 300 (which equals 4.9) would lead the installer to believe erroneously that five 110-type terminal blocks would be sufficient for terminating 370 4-pair cables. In fact, five 110-type terminal blocks are capable of terminating only 360 (5 × 72) 4-pair cables.

White - First-Level Backbone

The wall field sizing, block size, and block quantity are all based on the number of pairs contained in the selected first-level backbone cable. **See Figure 8-8.**

White fields are generally administered using 1-pair circuit modularity. Therefore, 110-C5 connecting blocks should be used when terminating backbone cables. When separate backbone cables exist for voice and data, it is recommended that voice and data cables be terminated on separate white fields. This makes administration much easier. It is usually more economical too, particularly when the voice equipment room and data equipment room are in separate locations.

Gray - Tie (Secondary Backbone) Cables

The wall field sizing, block size, and block quantity are all based on the number of pairs contained in the tie (secondary backbone) cables. Use the same sizing for tie cables as for first-level backbone cables.

Gray fields are generally administered using 1-pair circuit modularity. Therefore, 110-C5 connecting blocks should be used when terminating tie cables.

Purple - Common Equipment Circuits

For common equipment circuits terminated on 110-type terminal blocks, wall field sizing is based on two factors: quantity of circuits and pairs per circuit (circuit modularity). Circuit modularity is determined by the type of common equipment being supported and may range from as few as one pair to as many as four pairs per circuit. The block size and quantity will be determined by the total number of circuits to be terminated. **See Figure 8-9.**

Note: Except for 1-pair circuit modularity, the 25th pair on each row of 110-type hardware is not used.

To determine the number of rows of 110-type terminal blocks required, divide the total number of circuits by the quantity of circuits per row.

To determine the number of 110-type terminal blocks required, divide the total number of rows needed by the rows per block.

ARRANGING THE 110-TYPE TERMINAL BLOCK AT A TR WALL FIELD

The arrangement and space requirements for a 110-type wall field will depend on the type of hardware selected. The 110-type cross-connect system using terminal blocks requires about one square foot of wall space for each 300-pair 110-type terminal block. The 110-type panel system requires about two square feet of wall space for each 300-pair assembly.

To configure a 110-type terminal block TR wall field, follow these steps:

1. Determine the type of hardware to be used in the telecommunications rooms.

   110-type terminal blocks are for voice and low-speed data (10 Mb/s or less) and circuit administration is to be done using cross-connect wire.

   110-type panel kits are for data applications with speeds higher

Figure 8-8 Backbone Terminal Block Sizing

Backbone Cable Size	Recommended 110 Terminal Block Size
100-pair	One 100-pair
200-pair	One 300-pair
300-pair	One 300-pair
400-pair	One 300-pair and One 100-pair
600-pair	Two 300-pair

Figure 8-8. Technicians must be aware of the minimum terminal block size needed for multi-pair backbone cables.

Figure 8-9 Common Equipment Circuit Layout

Pairs-per-Circuit (Circuit Modularity)	Circuits-per-Row of 110 Blocks	110C Connecting Block Size
1-pair	25	5-pair
2-pair	12	4-pair
3-pair	8	3-pair
4-pair	6	4-pair

Figure 8-9. The number of circuits per row is determined by the common equipment pair requirements.

than 10 Mb/s, or if the customer requests circuit administration using 110-type patch cords.

Note: There are 110-type terminal blocks and panel kits rated for cables up to category 6A and data rates up to 10 Gb/s. The use of one of these systems for category 5e, 6, or 6A data rate applications may still be a desirable solution for facilities that may have high cable density installations. The application of 110-type blocks saves on space and on the cost of racking, especially in installations where there will not be a lot of moves, adds, or changes (MAC) work, such as in a data center.

2. List all circuits that will terminate on the wall field. For example:

   Horizontal

   Intra-building Backbone

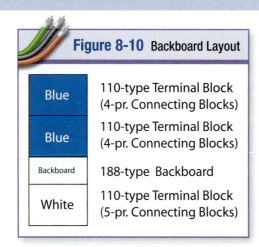

Figure 8-10 Backboard Layout

Blue	110-type Terminal Block (4-pr. Connecting Blocks)
Blue	110-type Terminal Block (4-pr. Connecting Blocks)
Backboard	188-type Backboard
White	110-type Terminal Block (5-pr. Connecting Blocks)

Figure 8-10. *A backboard typically separates cabling such as horizontal cabling and first-level backbone cables.*

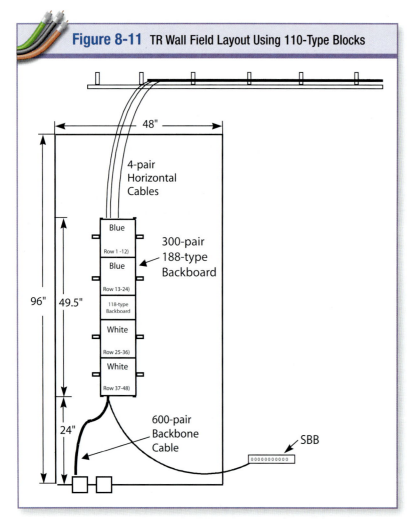

Figure 8-11 TR Wall Field Layout Using 110-Type Blocks

Figure 8-11. *Placing the backboard between the blue field and white field minimizes the length of the cross-connect cables.*

3. Determine quantity of each type of termination. For example:

   Ninety 4-pair horizontal cables, and one 300-pair intra-building backbone cable

4. Determine quantity of blocks per termination type. For example:

   Two 300-pair blocks for up to 144 4-pair horizontal cables, and one 300-pair block for one 300-pair intra-building backbone cable

5. Arrange blocks in a logical fashion and provide a backboard between fields.

6. On a drawing, position the block layout determined in Step 4 and then assign the proper label colors to the blocks. Finally, determine the circuit modularity for each type of termination. **See Figure 8-10.** For example:
   - 4-pair modularity (use 4-pair connecting blocks) for horizontal cables
   - 1-pair modularity (use 5-pair connecting blocks) for intra-building backbone cables

Although the arrangement of 110-type terminal blocks on a wall field is very flexible, these guidelines should be followed whenever possible.

Telecommunications Room 110-Type Wall Field

Consider a possible layout for a wall field in a telecommunications room utilizing 110-type terminal blocks. **See Figure 8-11.** The layout begins 24 inches above the floor line. This allows another block to be added below the backbone blocks if it is needed at a later time. The top of the wall field is six feet and 1½ inches high, which is the suggested maximum height.

The blue field is located above the 188-type backboard. The white field is below. The orientation of these fields may be dependent on the customer's preference. The recommendation, however, is to keep cable lengths as short as possible and provide a backboard in the center of the wall field to minimize the length of cross-connect wires.

110-Type Terminal Blocks with Patch Cords

Using 110-type patch cords is an acceptable practice; however, typical installations of 110-type terminal blocks usually do not provide adequate space for the routing of these patch cords. If the use of patch cords is anticipated, then adding troughs between 100-pair blocks can provide the necessary cord routing area. This will simulate 110-type panel kits. **See Figure 8-12.**

Consider a wall field for 110-type terminal blocks arranged to use patch cords. A 188-type backboard and troughs provide for routing of the patch cords between the different fields. **See Figure 8-13.** The recommendation is that an appropriately tall 188-type backboard be installed for every four columns of terminal blocks. In this example, however, there are only two columns of blue field in the center of the two 188-type backboards. Since both the backbone (white) and data equipment (purple) connect to the TOs, locating the blue field in the center reduces the length of the 110-type patch cords. The ports of the LAN switch in this room are cabled directly to the purple field. The blocks are located in columns A, B, C, and D with rows 1 to 12 from the top to the bottom.

110-Type Panel Kits

110-type panel kits may be used instead of 110-type terminal blocks and troughs. **See Figure 8-14.** 110-type panel kits use 100-pair cabling blocks that are similar to 110-type terminal blocks without legs. These cabling blocks are attached to a metal (or plastic) back panel. A trough, used for routing patch cords, is mounted above each 100-pair cabling block. A larger trough is mounted at the bottom of each vertical column of 110-type cabling blocks. A backboard is installed between vertical columns of 110-type cabling blocks. **See Figure 8-15.** The backboard is used for routing patch cords in the vertical plane. **See Figure 8-16.**

Figure 8-12 110-Type Patch Cords

Figure 8-12. 110-type patch cords may be used in some installations if adequate space is provided.

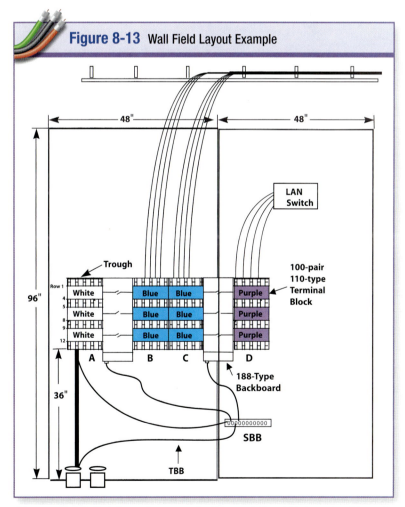

Figure 8-13 Wall Field Layout Example

Figure 8-13. Patch cords are routed through the 188-type backboard.

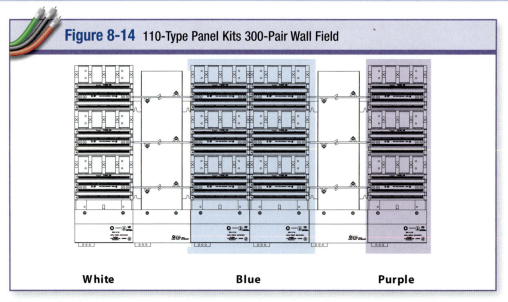

Figure 8-14 110-Type Panel Kits 300-Pair Wall Field

White   Blue   Purple

Figure 8-14. 110-type panel kits include vertical and possibly horizontal cable management built into the kit.

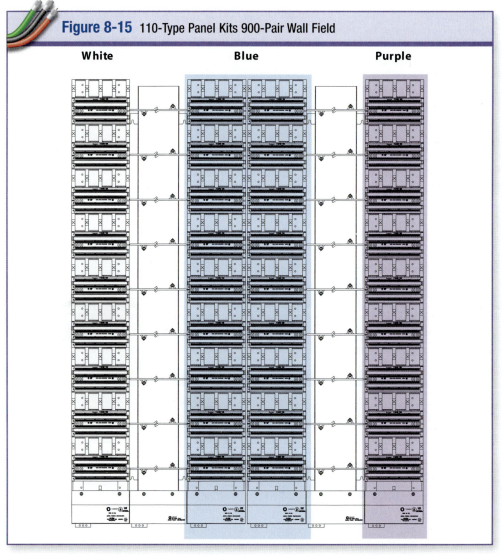

Figure 8-15 110-Type Panel Kits 900-Pair Wall Field

White   Blue   Purple

Figure 8-15. 110-type panel kits are modular and easily expandable.

CONFIGURING A RACK

Consider a typical equipment rack layout that may be found in a telecommunications room. **See Figure 8-17.** The 4-pair horizontal cables from work areas are terminated on the rear of the patch panel. This represents a blue field. The horizontal wire manager is provided to route and support modular patch cords used for interconnecting ports on the patch panel to ports on the LAN switch. In this example, the LAN switch has modular jack ports and represents a purple field. When common equipment is used to represent a purple field, it should be located as close as possible to the blue field hardware. The optional fiber patch panel shown would be present if there was a fiber backbone cable leaving the TR, destined for an ER. In that case, fiber patch cords would be used for connecting backbone fibers to fiber ports on the LAN hub.

In general, it is advisable to lay out racks so that patch panels and switches are alternated with a wire manager between them. **See Figure 8-18.** This minimizes the lengths of the cords used to connect the two fields and minimizes congestion in wire management. In a high-density environment, eliminating the horizontal wire management between patch panels and switches will save much needed room in the rack. Start by installing a patch panel and then a switch and then another patch panel, alternating between the two. Use one-foot long patch cables to connect the horizontal cables (blue field) with the LAN electronics (purple field). **See Figure 8-19.**

Configuring racks in equipment rooms is about the same as configuring a rack in a TR. The difference is that backbone cables are being terminated rather than horizontal cables.

CONFIGURING INTRA-BUILDING BACKBONE CABLING

The configuration steps for a UTP intra-building backbone cabling system are:

1. Determine voice and data backbone cabling requirements per TR.
2. Determine cable routing from each TR to each ER.
3. Determine the splicing method.

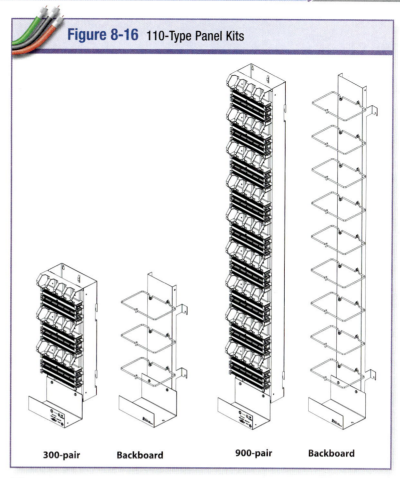

Figure 8-16 110-Type Panel Kits

300-pair   Backboard   900-pair   Backboard

Figure 8-16. *110-type panel kits may be used instead of termination blocks.*

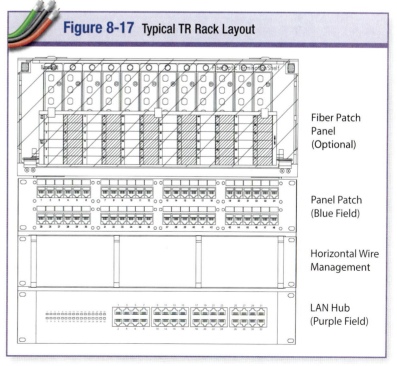

Figure 8-17 Typical TR Rack Layout

Fiber Patch Panel (Optional)

Panel Patch (Blue Field)

Horizontal Wire Management

LAN Hub (Purple Field)

Figure 8-17. *A TR rack layout may include a fiber shelf, patch panels, and horizontal wire management.*

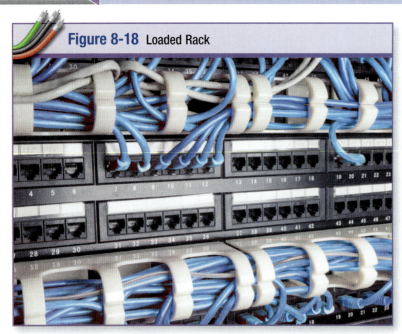

Figure 8-18 Loaded Rack

Figure 8-18. Horizontal wire managers between patch panels and/or switches are used to contain excess patch cable length and keep the rack components accessible.

Figure 8-19 High-Density

Figure 8-19. A rack in a high-density environment could save rack space by alternating patch panels with LAN electronics and using short patch cables between them.

Step 1: Determine Voice and Data Backbone Cabling Requirements per TR

When configuring twisted-pair backbone cabling, avoid the sharing of voice and digital data signals in the same cable sheath. When this is not feasible, utilize cables that have more than 25 pairs and contain individual 25-pair binder groups. In these cables, view each 25-pair binder group in the multi-pair cable as a separate 25-pair cable for sheath-sharing purposes. This way, different signal types may be transmitted within the same physical cable by isolating them within separate binder groups.

If the voice equipment and data equipment rooms are in different locations, configure separate voice and data backbone cabling systems to match the different routing requirements to their destinations.

Step 2: Determine Cable Routing from TRs to ERs

The ANSI/TIA-568.1-D Standard limits the cable distance from the main cross-connect (MC) to any TR to a maximum of 800 meters (2,624 ft) for analog voice applications. Category 5e and higher backbone cables intended for high-speed data applications are limited to a maximum length of 90 meters (295 ft). If the distance from the MC to any TR exceeds this maximum limit, an intermediate cross-connect (IC) must be established. In these cases, the maximum cable length of 90 meters will apply to the MC-to-IC cable and to the IC-to-TR cable. A LAN hub or repeater will usually be required at the IC for the purpose of regenerating the signal as it passes through.

Step 3: Determine the Splicing Method

In designing the backbone, the installer must decide what splicing strategies to use. There are three major ways to run backbone cabling to termination hardware on a given floor:

1. Point-to-point termination (home run)
2. Branch splice
3. Intermediate cross-connect

Point-to-Point Termination

Point-to-point termination is the simplest and most direct method. It involves selecting a cable that has a sufficient number of twisted pairs or fibers to support the communication needs for just one telecommunications room on one floor. That cable is then run from the equipment room through the backbone pathway to the TR. A cable's length is a direct function of which floor it has been assigned to and of the distance from the TR to the MC. The major advantages of point-to-point termination are that it allows for smaller, less bulky, and more flexible cables and it eliminates the need for expensive splicing.

A disadvantage is that a greater number of cables need to be pulled through the backbone pathway. **See Figure 8-20.**

Branch Splice

When using branch splices, a large feeder cable with the capacity to support users on several floors is run within the backbone pathway system. On selected floors, pairs from the large cable are spliced to pairs in smaller ones, which are routed from the splice closure to other TRs, where they are terminated. **See Figure 8-21.**

The advantage of branch splicing is that overall, fewer backbone cables are needed, which may save some space. Under certain conditions, branch splicing is sometimes less expensive than point-to-point terminations. A decision as to which of the two major approaches is best for a building is usually a trade-off between the cost of cable and cost of the labor needed for splicing.

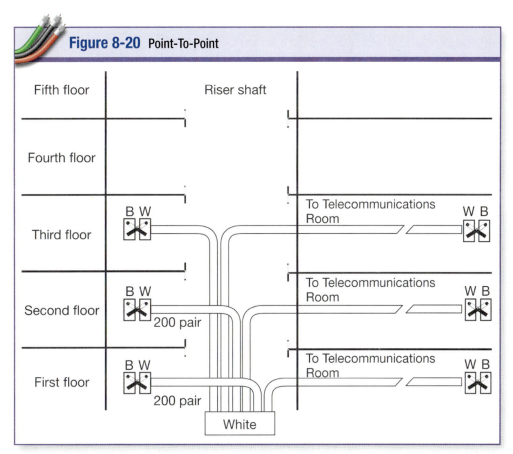

Figure 8-20 Point-To-Point

Figure 8-20. 200-pair cables can be routed point-to-point through the backbone pathway to individual TRs throughout a building.

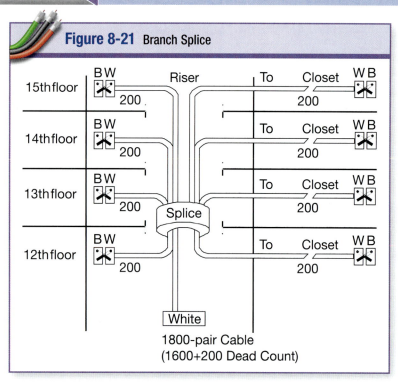

Figure 8-21 Branch Splice

Figure 8-21. An 1,800-pair cable spliced to support four floors in a multi-story building is an example of a typical branch splice layout.

Intermediate Cross-Connect

When there are multiple TRs on a floor, it is sometimes desirable to administer all of the horizontal cables from a single location. In this case, second-level backbone (tie) cables are installed between the first-level TR and the other TRs on the floor. The backbone TR now houses an intermediate cross-connect (IC), which serves as the primary administration point for the entire floor.

At an IC, pairs from a first-level backbone cable terminated on a white field are cross-connected to pairs in second-level backbone (tie) cables, which are terminated on a gray field. At a remote TR, pairs from a second-level backbone (tie) cable, terminated on a gray field, are cross-connected to pairs in horizontal cables, which are terminated on a blue field. These cross-connections are usually done upon initial installation and are generally permanent. **See Figure 8-22.**

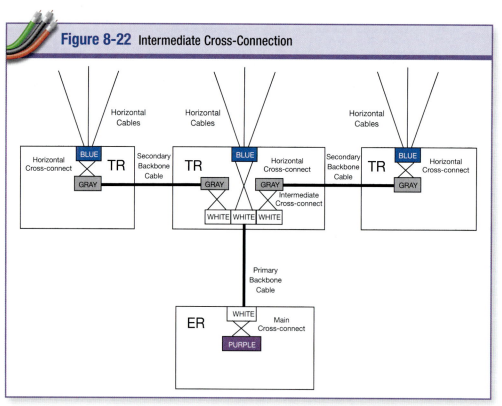

Figure 8-22 Intermediate Cross-Connection

Figure 8-22. An intermediate cross-connection for a second level backbone (tie) cable creates a single point of administration for several TRs on the same floor.

CONFIGURING A WALL FIELD IN AN EQUIPMENT ROOM

Consider an equipment room wall field layout utilizing 110-type terminal blocks. **See Figure 8-23.** The purple field begins 12 inches above the floor line, the minimum recommended distance from the floor for a wall field. The top of the wall field is six feet and 1½ inches high, the suggested maximum height. There are six columns of 110-type terminal blocks. A two-inch space is left between the columns to provide space to add distribution (D) rings for routing jumper wires.

The green, white, brown, and blue fields are located above the row of backboards. The purple field is located below. The orientation of these fields may be dependent on the customer's preference, but the recommendation is to keep cable lengths as short as possible. Provide a row of backboards in the center of the wall field to minimize the length of cross-connect wires.

Arrange the circuit terminations so that pairs terminated on the top of the field will cross-connect to equipment port appearances on the bottom of the field. Grow the wall field horizontally in order to minimize the height.

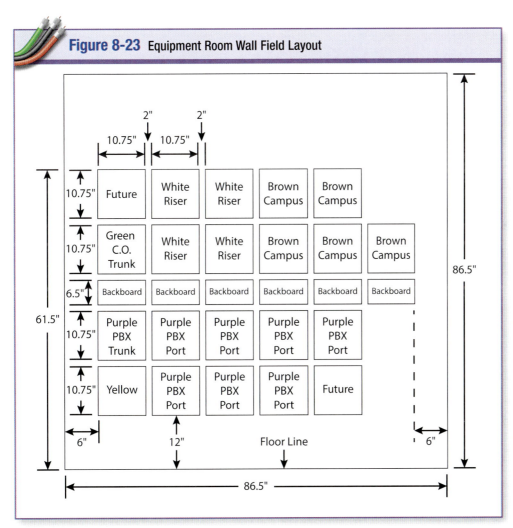

Figure 8-23 Equipment Room Wall Field Layout

Figure 8-23. *When arranging the ER wall layout, minimize cable and cross-connect lengths as much as possible.*

SUMMARY

The configuration of a structured cabling system relies on a competent design and careful consideration of how terminations are made. There are a number of factors to be considered, including distance for horizontal cabling, average cable lengths, and how a wall field is configured. A color code is utilized on the wall field to identify termination connections. A technician must understand the number of pairs that can be connected to a 110-type terminal block as well as how many 110-type blocks are needed for a structured cabling system.

REVIEW QUESTIONS

1. How many steps must be followed to determine how much of each type of horizontal cable is required for a job?
   a. 3
   b. 4
   c. 5
   d. 6

2. What wall field color is typically used for first-level backbone cables between the MC and a TR?
   a. Blue
   b. Gray
   c. Purple
   d. White

3. The typical spool size for 4-pair UTP cable is 500 feet.
   a. True
   b. False

4. How many 4-pair circuits per row can be achieved on a 100-pair 110-type terminal block?
   a. 6
   b. 8
   c. 12
   d. 25

5. What is the recommended 110-type terminal block size for a 300-pair cable?
   a. One 100-pair and one 200-pair
   b. One 300-pair
   c. Three 100-pair
   d. Any of the above

6. There are __?__ major ways to run backbone cabling to termination hardware on a given floor.
   a. 2
   b. 3
   c. 4
   d. 5

7. At an intermediate cross-connect, the second-level backbone cables are terminated on a __?__ field.
   a. blue
   b. gray
   c. purple
   d. white

8. What is the minimum recommended distance above the floor to begin a wall field?
   a. 12"
   b. 18"
   c. 24"
   d. 48"

9. The disadvantage of using point-to-point termination with backbone cabling is the quantity of cables being pulled through the backbone pathway.
   a. True
   b. False

Residential Cabling Systems

Introduction

There are a number of items that must be considered when configuring and installing a residential structured cabling system: where the distribution device will be installed, where the DMARC will be installed, what systems will be installed, what cable types and grades will be needed, where the outlets will be located, and much more. This chapter will introduce several new terms, explain the different parts of a residential phone system and their purposes, and recommend ways to install audio and video systems.

Objectives

- Utilize the ANSI/TIA-570-D *Standard for Residential Telecommunications Infrastructures*.
- Explain the differences between the grades of residential cabling and where they should be implemented.
- Explain a NID, an ADO, and a DMARC.
- Plan the installation of a whole house audio system and video system.

Chapter 9

Table of Contents

TYPICAL RESIDENTIAL CABLING

In a typical residential dwelling, each system (phone, television, HVAC, security, sound, etc.) has its own cabling, and the type of cable used by each system is unique. Phone wiring in several existing dwellings may still consist of the older style quad wire; however, current standards require category 6A UTP to be installed in all renovations and new construction of single- and multi-dwelling residential buildings. Television wiring is typically 75-ohm coaxial cable. The current standards recommend, at a minimum, the use of series-6 coaxial cable (RG-6). HVAC and security systems utilize multi-conductor cables (sometimes solid and sometimes stranded conductors), and in some cases category 3 through category 6A UTP is used to connect the devices used in these systems. Sound systems typically employ heavier-gauge stranded wires. **See Figure 9-1.**

TRADITIONAL TELEPHONE WIRING

Older styles of residential telephone wiring consisted of either quad wire or twisted-pair wire. Quad wire (also known as POTS wire for "Plain Old Telephone Service") is made up of four insulated conductors that are not twisted around one another. The red and green wires make up a pair, which is used for line 1. The yellow and black wires make up a second pair, used for line 2. **See Figure 9-2.**

Twisted-pair telephone wire comes in pair counts from 2- to 6-pair. The white/blue pair is used for line 1, and the white/orange pair is used for line 2.

The use of legacy types of telephone wiring were eliminated by a January 10, 2000 FCC ruling that set new minimum quality standards for telephone inside wiring using category 3 cable. Title 47 of the Code of Federal Regulation (CFR), Part 68.213, *Installation of other than "fully protected" non-system simple customer premises wiring,* part (c) states:

(1) For new installations and modifications to existing installations, copper conductors shall be, at a minimum, solid, 24 gauge or larger, twisted pairs that comply with the electrical specifications for Category 3, as defined in the ANSI/TIA/EIA Building Wiring Standards.

Effective 180 days after the FCC ruling, new POTS wire installed in a resi-

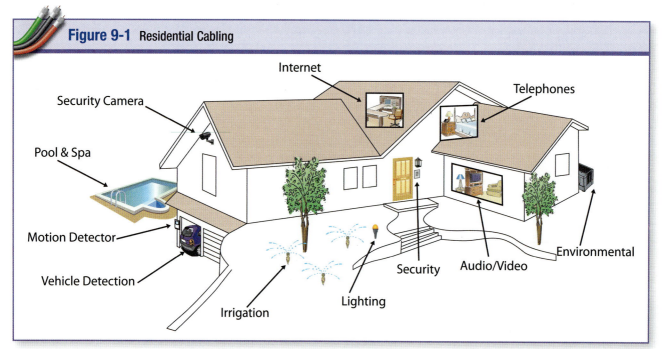

Figure 9-1 Residential Cabling

Internet

Telephones

Security Camera

Pool & Spa

Motion Detector

Vehicle Detection

Irrigation

Lighting

Security

Audio/Video

Environmental

***Figure 9-1.** Modern homes contain wiring for multiple different systems.*

dence for telecommunications use was considered illegal.

RESIDENTIAL TELECOMMUNICATIONS CABLING STANDARD

Because of the widespread acceptance of the ANSI/TIA-568.1 *Commercial Building Telecommunications Cabling Standard,* the TIA decided to develop a similar wiring standard for residential cabling. The ANSI/TIA/EIA-570-B *Residential Telecommunications Infrastructure Standard* (April 2004) defined the requirements for residential telecommunications cabling. These requirements were based on the facilities that are necessary for existing and emerging telecommunications services. The cabling infrastructure specifications within the TIA-570-B Standard were intended to include support for voice, data, video, multimedia, home automation systems, environmental control, security, audio, television, sensors, alarms, and intercom. The TIA-570-B Standard was intended to be implemented for new construction, additions, and remodeled single- and multi-tenant residential buildings. This Standard also applies to the telecommunications cabling within or between structures and includes the cabling within a single-dwelling unit and the backbone cabling.

This ANSI/TIA/EIA-570-B Standard replaced ANSI/TIA/EIA-570-A, published in September 1999. The ANSI/TIA/EIA-570-B Standard incorporated and refined the technical content of:

- ANSI/TIA/EIA-570-A-1, *Residential Telecommunications Cabling Standard; Addendum 1 – Security Cabling for Residences*
- ANSI/TIA/EIA-570-A-2, *Residential Telecommunications Cabling Standard; Addendum 2 – Control Cabling for Residences*
- ANSI/TIA/EIA-570-A-3, *Residential Telecommunications Cabling Standard; Addendum 3 – Whole-Home Audio Cabling for Residences*

The incorporation and refinement of these cabling references and their inclu-

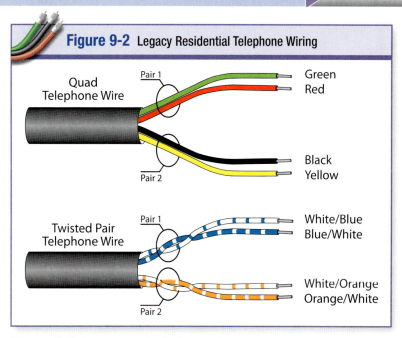

Figure 9-2   Legacy Residential Telephone Wiring

Figure 9-2. Legacy quad telephone wire was susceptible to interference due to the lack of twisting of the conductors.

sion into the ANSI/TIA/EIA-570-B Standard basically standardized types of cabling that have been commonly installed in residences for years.

In August 2012, TIA released ANSI/TIA-570-C. This standard incorporated and refined the technical content of ANSI/TIA/EIA-570-B and Addendum 1. In July 2018, TIA released ANSI/TIA-570-D, which replaced ANSI/TIA 570-C. Changes from the previous document include:

- Addition of Grade 3
- Replacement of the term "telecommunications outlet/connector" with "equipment outlet"
- Specification of category 6A as the minimum for balanced twisted-pair cabling
- Reduction of required coaxial cables from two to one
- Removal of series-59 coaxial cable (RG-59) as a recognized medium
- Removal of the 100-meter (328 ft) length restriction for optical fiber outlet cabling
- Other minor changes

NEC® REQUIREMENTS

The 2008 *National Electrical Code* added a requirement to Chapter 8 mandating that any new dwelling construction have a minimum of at least one communications outlet installed. Section 800.156, Dwelling Unit Communications Outlet, states:

> A minimum of one communications outlet shall be installed within the dwelling and cabled to the service provider demarcation point.

The *NEC®* made no attempt to provide a location for the communications outlet, nor did it provide for minimum media requirements. However, Chapter 8 attempts to provide guidance to "acceptable industry practices" through an Informational Note under Article 800.24, Mechanical Execution of Work. This Informational Note makes reference to ANSI/TIA-570, *Residential Telecommunications Infrastructure Standard,* as well as other industry recognized standards.

In the 2020 *NEC,* Dwelling Unit Communications Outlet was moved to Section 805.156, and a requirement was added mandating placement to be in a "readily accessible area."

GRADES OF RESIDENTIAL CABLING
Within the TIA-570-D Standard, there are three grades of cabling defined for residential units: Grade 1, Grade 2, and Grade 3. A grading system was established in the TIA-570-A Standard based upon services that are expected to be supported within each residential unit and to assist in the selection of the cabling infrastructure. For home automation cabling requirements, the Standard refers the installer to the manufacturer's recommendations.

Grade 1
For each cabled location, Grade 1 provides a generic cabling system that meets the minimum requirements for telecommunications services. For example, this grade provides for telephone, satellite, community antenna television (CATV), and data services. Grade 1 specifies balanced twisted-pair cabling and coaxial cabling. Grade 1 cabling minimum requirements are for one 4-pair UTP cable

that meets or exceeds the requirements for category 6A cabling, and include a minimum of one 75-ohm coaxial cable and the respective connectors at each equipment outlet (EO) and distribution device (DD). Optionally, 2-fiber (minimum) optical fiber cabling may be deployed in addition to the twisted-pair and broadband coaxial cabling.

Grade 2
For each cabled location, Grade 2 provides a generic cabling system that meets the requirements for basic and advanced telecommunications services. This grade provides for both current and developing telecommunications services. Grade 2 specifies balanced twisted-pair cabling and coaxial cabling. Grade 2 cabling minimum requirements are for two 4-pair UTP cables and associated connectors that meet or exceed the requirements for category 6A cabling; in addition, Grade 2 requires one 75-ohm coaxial cable and associated connectors at each equipment outlet and the distribution device; optionally, 2-fiber optical fiber cables may also be deployed.

Grade 3
For each cabled location, Grade 3 provides a generic cabling system that meets the minimum requirements for basic and advanced telecommunications services such as high-speed Internet, wireless access points, and in-home generated video. This grade provides for both current and developing telecommunications services. Grade 3 specifies balanced twisted-pair cabling, coaxial cabling, and optical fiber cabling. Grade 3 cabling requirements consist of a minimum of two 4-pair balanced twisted-pair cables that meet or exceed the requirements for category 6A, a minimum of one broadband coaxial cable, and a minimum of one 2-fiber optical fiber cable along with their respective connectors at each equipment outlet and the distribution device.

RESIDENTIAL CABLING ARCHITECTURE
The demarcation point (DMARC) is the interface point between the access provider and customer facilities. The demar-

cation point may be evidenced by a network interface device (NID) provided and installed by the access provider. For single-family residences, the demarcation point is usually located on the outside of an exterior building wall.

In the event the NID has not been installed, the access provider shall be contacted to locate the demarcation point according to applicable regulations. The telephone service provider's drop wire is terminated in the NID. **See Figure 9-3.** After electrical (overvoltage) protection is applied, the pairs are terminated on modular jacks in the NID. The auxiliary disconnect outlet (ADO) cable plugs into the NID and extends the circuits to the ADO. Because the ADO provides the means for the tenant to disconnect from an access provider, the ADO device can be used for verifying whether the access provider signal has been dropped or if there is an internal wiring problem. **See Figure 9-4.** This is accomplished by taking a phone from the dwelling or a lineman's butt set and plugging it into the jack. If the phone operates, the fault is in the residential wiring. If the phone does not operate, the access provider must be contacted to start a repair ticket.

It is desirable to co-locate the ADO with the distribution device (DD). The Standard states "the ADO and the DD shall be located indoors and be readily accessible."

ADO cables extend services from the demarcation point to the ADO. If the single residential unit is part of a multi-residential building, the ADO cables may extend from the common telecommunications room (CTR) to the ADO located in the tenant space. **See Figure 9-5.**

A DD shall be provided within each residence. The DD is a cross-connect facility used for the termination and connection of outlet cables, DD cords, equipment cords, and in some cases ADO cables. The DD is used for connection of access providers to the residence and to facilitate moves, adds, and changes of premises cabling within the residence. Access to the building electrical ground shall be provided within 1.5 meters (5 ft) of the DD, in accordance with applicable codes.

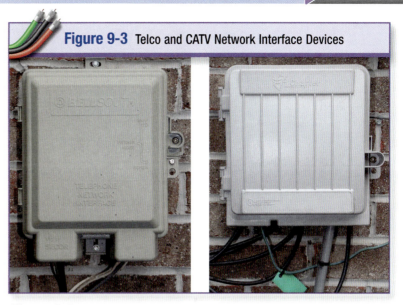

Figure 9-3 Telco and CATV Network Interface Devices

Figure 9-3. The telco NID (left) and the CATV NID (right) are typically called the demarcation point, or DMARC. The connection between the customer premises cabling and the service provider's (SP) cabling happen here.

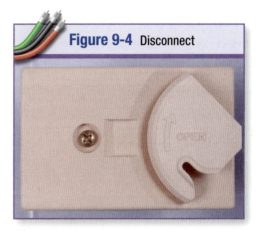

Figure 9-4 Disconnect

Figure 9-4. The auxiliary disconnect outlet is the point where the premise cabling may be disconnected from the service provider cabling.

The DD may be designed to distribute any and all of the following: audio/video (A/V), whole house or zone audio, CATV, CCTV, security, and home automation control (HVAC, lighting, irrigation, etc.). **See Figure 9-6.**

Note: distribution devices may be known as structured media centers (SMC), media convergence centers, etc.

Outlet cables provide the transmission path from the DD to the outlet. The length

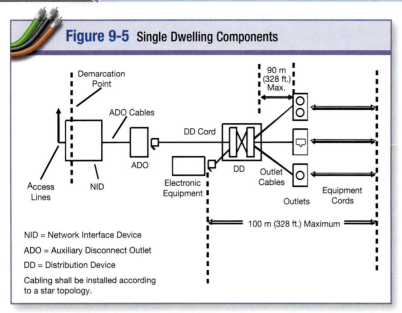

Figure 9-5 Single Dwelling Components

NID = Network Interface Device

ADO = Auxiliary Disconnect Outlet

DD = Distribution Device

Cabling shall be installed according to a star topology.

Figure 9-5. The ADO is the point where the tenant may disconnect from an access provider's service.

Figure 9-6 Distribution Device

Figure 9-6. Leviton offers a structured media center with an ADO and voice, data, and video connections in one package.

of each outlet cable shall not exceed 90 meters (295 ft). The 90-meter length allows an operational length of 100 meters (328 ft), including patch cords or jumper wire and equipment cords.

Recognized outlet cables include:

- 4-pair 100-ohm UTP (category 6A)
- Series-6 coax (commonly known as RG-6), dual- (also called single tape and braid cable), tri-, or quad-shield
- Series-11 coax (commonly known as RG-11)
- 2-fiber (minimum) optical fiber (OM4 or OM5 recommended)

To use multiple hard-line analog phones throughout the dwelling, a bridging block may be constructed using a 110-type terminal block and wired as follows: beginning at the top row, paired cross-connect wire is used to multiple (combine) every first-pair position, beginning with position 1. The result will be the interconnection of positions 1, 5, 9, 13, 17, and 21. The D-impact tool must be set to seat, but not cut, the wire. Using a second-paired cross-connect wire, multiple (combine) every second-pair position, beginning with position 2.

The result will be the interconnection of positions 2, 6, 10, 14, 18, and 22.

This arrangement provides the ability to multiply the appearance of a single telephone line up to six times on a single row of a 110-type terminal block. An additional row on the 110-type terminal block is wired in the same manner to provide for the second phone line. Successive rows are used for additional phone lines. **See Figure 9-7.**

Per TIA-570-D, a single dwelling residence shall include a minimum of one outlet location be cabled within each of the following rooms (where applicable):

- Kitchen
- Each bedroom
- Family/great room
- Office/study
- Home theater

In addition, a sufficient number of telecommunications outlet locations should be planned to prevent the need for extension cords. An outlet location should be provided in each room and additional outlet locations provided within unbroken wall spaces of 3.7 meters (12 ft) or more. Additional outlet locations should be provided so that no

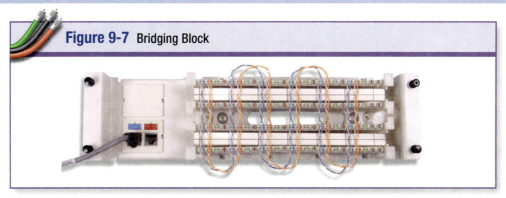

Figure 9-7 Bridging Block

Figure 9-7. *A bridging block may be constructed using a 110-type terminal block and wired as shown. This allows for multiple appearances of a single line.*

point along the floor line in any wall space is more than 7.6 meters (25 ft) from an outlet location in that space. **See Figure 9-8.**

UTP cables shall be terminated on 8-position/8-conductor outlets using the T568A configuration.

Coaxial cable is used for satellite, CATV, and CCTV systems. Satellite and CATV systems shall use series-6 outlet cable in lengths up to 46 meters (150 ft). Series-11 or hard-line trunk shall be used for coaxial backbone cabling when needed.

Series-6 coaxial cables shall meet the requirements of ANSI/SCTE 74 or the Structural Building Components Association Series 6 Recommended Practices.

Series-11 coaxial cables shall meet the requirements of ANSI/SCTE 74.

Trunk, feeder, and distribution cables shall meet the requirements of ANSI/SCTE 15.

Braided multipurpose cables shall meet the requirements of ANSI/SCTE 71.

MULTI-DWELLING UNIT BASIC DESIGN CONSIDERATIONS AND RECOMMENDATIONS

1. A minimum of one outlet location should be cabled in each of the following rooms:
   - Kitchen
   - Bedroom
   - Family/great room
   - Office/study
   - Home theater
2. Outlet cables shall be home run from outlets to the DD.

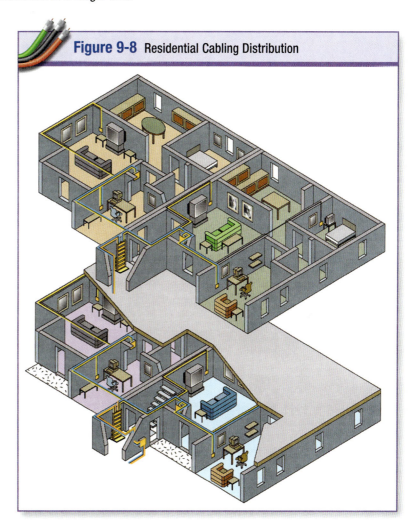

Figure 9-8 Residential Cabling Distribution

Figure 9-8. *Outlet locations need to be planned for each room.*

3. The DD should be centrally located to minimize cable run lengths and be in a climate-controlled location.

4. Minimum recommended back-bone cables to the DD:
   - 4-pair or multi-pair balanced twisted-pair cable
   - 75-ohm series-6 or -11 coaxial cable
   - Optical fiber, OM4 or OM 5 recommended
5. A 120-volt AC, 15-ampere non-switchable duplex electrical outlet should be provided within the DD.

DISTRIBUTED AUDIO

In a distributed audio system, audio is distributed and controlled over UTP using an A-BUS system.

Line-level audio, IR, and source on/off status is distributed from a central source (for example, PC, TV, CD player, etc.) over category 5e or higher cabling to amplified volume control modules located in each room, thus eliminating the need for a separate amplifier.

Any category 5e/6/6A outlet jack can be used for audio distribution by simply connecting the source input module using a standard category 5e patch cable, then making the appropriate connection in the DD. Speaker cable (16 AWG recommended) is only required between the volume control and speaker pair located in each room. Volume controls are available with or without IR capability. **See Figure 9-9.**

A-BUS distribution hubs support up to four zones. The number of zones can

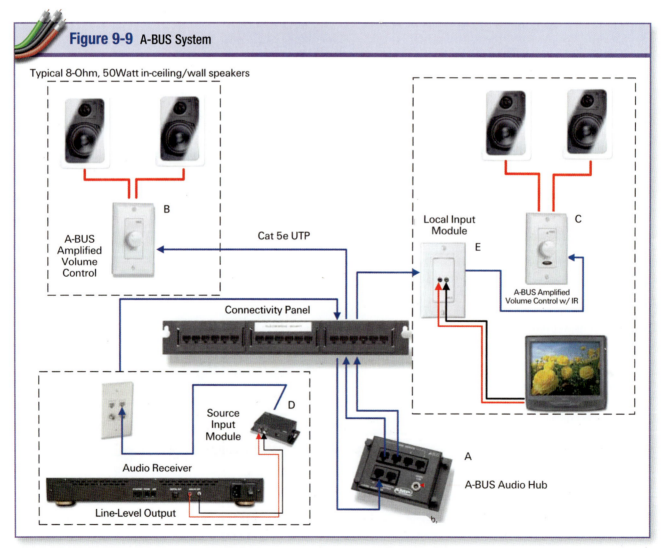

Figure 9-9 A-BUS System

Typical 8-Ohm, 50Watt in-ceiling/wall speakers

B — A-BUS Amplified Volume Control

Cat 5e UTP

Connectivity Panel

Source Input Module — D

Audio Receiver

Line-Level Output

A — A-BUS Audio Hub

Local Input Module — E

C — A-BUS Amplified Volume Control w/ IR

Figure 9-9. *A basic whole house audio system can be installed using A-BUS technology.* Courtesy of Tyco Electronics *(A-BUS is a trademark of Leisuretech Electronics Pty Ltd.)*

Figure 9-10 A-BUS Components

4-Port Hub Kit w/Power Supply   *Amplified Volume Control*   *Amplified Volume Control w/IR*

Source Input Module   *Local Input Module*

Figure 9-10. Many different components make up the A-BUS system.

be expanded by adding additional hubs via the expansion port. Local input modules allow each audio zone to be used for independent listening. **See Figure 9-10.**

VIDEO DISTRIBUTION

In a typical video distribution system, CATV and/or off-air antenna broadband video signals are combined and distributed along with CCTV, satellite (DBS), DVD/BluRay, and digital video recordings (DVR) throughout the dwelling. **See Figure 9-11.**

In some instances, off-air signals or CATV signals coming into the home are combined with an A/V source (this could be a central DVD changer, for example). **See Figure 9-12.** The signals are then distributed through the home using a combiner and a 4-way splitter. In this example, signal loss through the combiner is 3.5 decibels. There is another seven-decibel loss from passing through the 4-way splitter. Every time a pass-through device or splitter is added, the video signal is reduced.

Notice that the cable lengths vary as they travel to the different television sets. Losses in the coax are based on the type of coax used and its length. Because of the short runs to the first and second TVs (from the right) the signals would be to "hot" for the front end (tuner) of the TV. Ideally, there should be between 10 and 12 decibels (and in no case higher than 15 decibels) of video signal at the TV for

Figure 9-11 Basic Video System

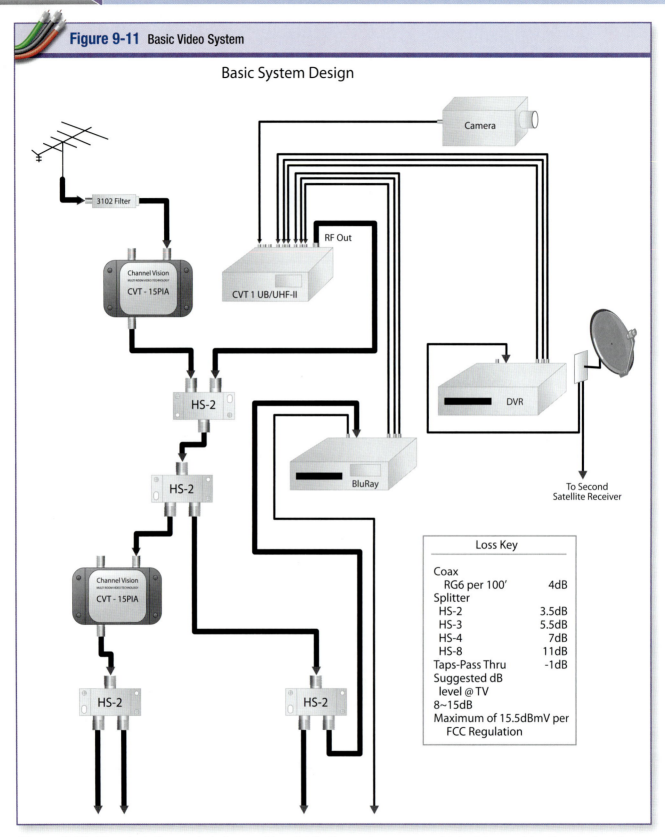

Figure 9-11. When installing a whole-house video distribution system, one must take into consideration the loss from cable resistance and the components used. Courtesy of Channel Vision

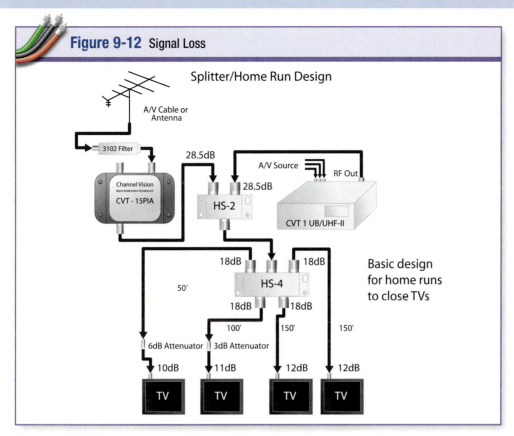

Figure 9-12 Signal Loss

Figure 9-12. *When running cabling for video, adjustments for signal strength may need to be made. Splitters and attenuators may be used to reduce decibel level at the television when needed. Courtesy of Channel Vision*

best picture quality. Because of the high signal level being supplied to the first two TV sets, attenuators need to be added to reduce the video signals to acceptable levels.

To maintain signal quality and to reduce standing waves (impedance mismatches) that will degrade picture quality, all unused video outlets should be terminated using 75-ohm terminators.

SUMMARY

There are many different parts to a residential system. The NID, ADO, and DMARC all work together for proper functioning of a residential system. There are several grades of cabling, and the ANSI/TIA residential standard suggests locations for each. The standards should be referenced when installing distributed audio and video systems to get the best result.

REVIEW QUESTIONS

1. Older styles of residential quad wire were used for POTS lines. The insulated pairs were twisted around one another.
   a. True
   b. False

2. Which ANSI/TIA standard is the standard for the residential telecommunications infrastructure?
   a. ANSI/TIA-568
   b. ANSI/TIA-569
   c. ANSI/TIA-570
   d. ANSI/TIA-606

3. How many grades of residential cabling are there according to the standards?
   a. 1
   b. 2
   c. 3
   d. 4

4. Which grade of cabling meets the minimum requirements?
   a. Grade 1
   b. Grade 2
   c. Grade 3
   d. Grade 4

5. The standard states "the ADO and the DD shall be located __?__ and be readily accessible."
   a. in the basement
   b. indoors
   c. outdoors
   d. out of sight

6. Which of the following is not a recognized outlet cable?
   a. Optical fiber OM1 or OM2
   b. Optical fiber OM 4 or OM5
   c. Series-6
   d. UTP

7. What provides the means for a customer to disconnect from an access provider?
   a. ADO
   b. DD
   c. DMARC
   d. NID

8. Which wiring standard is recommended for an 8-position/8-conductor modular jack in a residential application?
   a. T568A
   b. T568B

9. In a distributed audio system, what size speaker cable is recommended between the volume control and speaker pair?
   a. 14 AWG
   b. 16 AWG
   c. 18 AWG
   d. 20 AWG

10. Within a video distribution system, every time a pass-through device or splitter is added, the video signal __?__.
    a. decreases
    b. increases
    c. stays the same

Testing Structured Cabling Systems

Introduction

After all of the planning, cable installation, terminations, and administration tasks are complete, it is time to make sure everything will work properly for the customer; it is time to test the cabling system. A quick look in the job specifications will indicate the type of testing that will need to be performed, and that type of testing will indicate exactly what measurements will be made. Will it be verification, qualification, or certification testing? What kind of links will be tested: permanent, channel, or modular plug terminated links?

This chapter will explain the different types of testing and the parameters verified while performing those tests.

Objectives

- Describe certification, qualification, and verification testing.
- List and explain the different tests performed by the certification tester.
- Explain the difference between permanent link, channel link, and modular plug terminated link testing.

Chapter 10

Table of Contents

TESTING A STRUCTURED CABLING SYSTEM

Whether the cabling is used for voice and data or for building automation and control, the structured cabling system is the foundation of a building's network. Removing and/or replacing the cabling is expensive and time-consuming, not to mention very disruptive to the customer's business operations. This is even more true when removal or replacement is done after the cabling system has been commissioned.

The cabling system will outlast the active electronics that are connected to it; designing and certifying the cabling system for future changes in technology is critical. To ensure that the cabling system will perform as required, now and in the future, the cabling system and hardware must be installed following the manufacturer's recommendations and the industry standards. Then, the system must be tested for proper performance.

LEVELS OF FIELD TESTING

Some type of performance testing is usually specified or required for a new cabling system. Testing of the cable requires that it be measured against a known performance requirement for that category of cable.

CERTIFICATION FIELD TESTING

Cable performance is defined by industry standards, and these performance specifications can be found in the ANSI/TIA-568.2-D, *Balanced Twisted-Pair Telecommunications Cabling and Components Standard* (Section 6) and the ISO Standard 11801-1.

Certification testing assures that the cabling link meets all of the transmission performance requirements, including the bandwidth of the cable and its signaling transmission capacity. **See Figure 10-1.**

With a corresponding increase in the signaling transmission of a cable (that is, category 5e cabling at 100 MHz or 1 Gb/s up to category 6A at 500 MHz or 10 Gb/s), more data can be pushed through it.

> **REMINDER**
>
> Bandwidth and data rates are related, but they are not the same thing. **See Chapter 4 for more information.**

Certification of cable performance requires that the test be documented and reported for future evaluation and/or inspection. Documentation and reporting is done by, and stored in, the certification test tool. The cabling performance information recorded in the tester is typically submitted to the designers/consultants, the owner, and the cabling system manufacturers for warranty programs. **See Figure 10-2.**

Cable certifier measurement accuracy is independently verified by a third-party testing lab to comply with levels defined by the ANSI/TIA-1152-A and the ISO/IEC 61935-1 Ed 5. The accuracy calibration is required on a yearly basis to ensure accurate measurements. **See Figure 10-3.**

The signaling applications table provides some examples of various types of signaling applications, the supported cabling, and the corresponding data rates and bandwidth. It is important to note that the bandwidth and data rates shown

Figure 10-1 IDEAL Networks LanTEK IV Cable Certifier

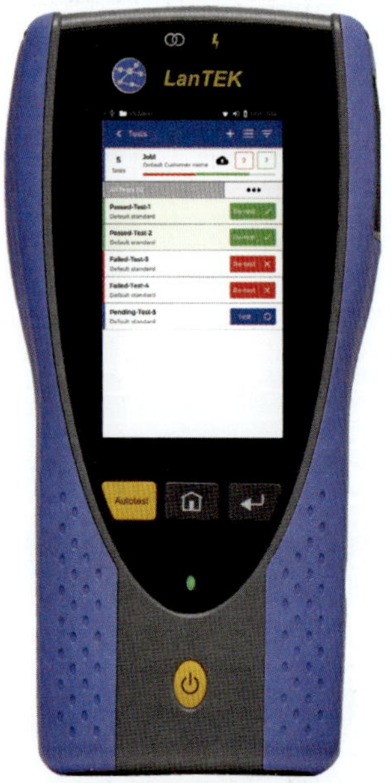

Figure 10-1. Certification cable testers must have warranty submissible reporting per TIA/ISO Industry Standards. Courtesy of IDEAL Networks

Figure 10-2 Test Report for *electrical training ALLIANCE* Office

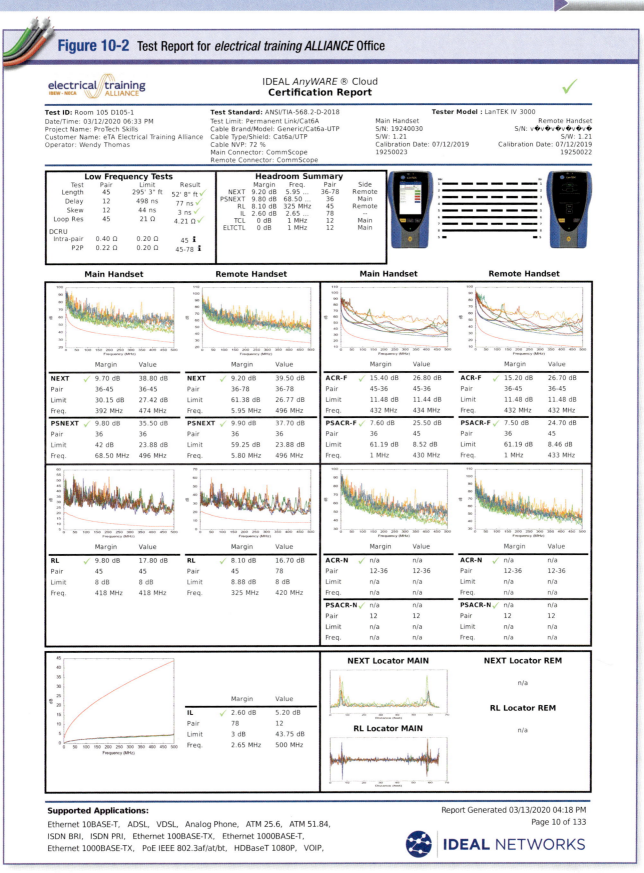

Figure 10-2. *A test report from the LanTEK IV certification test tool for an office data cable can be created for submission to the owner and cable manufacturer for warranty.*

Figure 10-3 Certifier Accuracy

Accuracy Level	Accrediting Body/ Standards Development Organization	Max Testing Frequency	Cabling Standard (TIA/IEC)	Supported Connectors	Supported Ethernet Data Rate
Level IIe	ANSI/TIA & ISO/IEC	100 MHz	Cat 5e/Class D	RJ45	1 Gb/s
Level III	ANSI/TIA & ISO/IEC	250 MHz	Cat 6/Class E	RJ45	1 Gb/s
Level IIIe	ANSI/TIA & ISO/IEC	500 MHz	Cat 6A/Class Ea	RJ45	10 Gb/s
Level IV	ISO/IEC	600 MHz	Class F	TERA/GG45/EC7	25 Gb/s
Level V	ISO/IEC	1000 MHz	Class Fa	TERA/GG45/EC7	25-40 Gb/s
Level VI	ISO/IEC	2000 MHz	Class I	RJ45	40 Gb/s
Level VI	ISO/IEC	2000 MHz	Class II	TERA/GG45/EC7	40 Gb/s
Level 2G	ANSI/TIA	2000 MHz	Cat 8.1	RJ45	40 Gb/s
Level 2G	ANSI/TIA	2000 MHz	Cat 8.2	TERA/GG45/EC7	40 Gb/s

Figure 10-3. Manufacturers require their test sets be calibrated once a year to guarantee accuracy in certification.

Figure 10-4 Signaling Applications

Application			TIA/Passive Cabling	
Application	Data Rate (excluding overhead)	Bandwidth	Minimum Cable Specification	Cabling Bandwidth
10Base-T	10 Mb/s	7.5 MHz	Category 3	16 MHz
1000Base-T	1,000 Mb/s	25 MHz	Category 5/5e	100 MHz
1000Base-Tx	1,000 Mb/s	250 MHz	Category 6	250 MHz
10GBase-T	10,000 Mb/s	500 MHz	Category 6A/7	500/600+ MHz
2.5GBase-T	2,500 Mb/s	100 MHz	Category 5e	100 MHz
5GBase-T	5,000 Mb/s	200 MHz	Category 6	250 MHz
40GBase-T	40,000 Mb/s	1,600 MHz	Category 8	2,000 MHz

Figure 10-4. The application data rates and bandwidth must be supported over the entire link.

in the table are supported over the entire 100-meter (328 ft) channel length. **See Figure 10-4.**

QUALIFICATION FIELD TESTING

Qualification field testing assures that the installed cabling link will successfully transmit data using a specific network technology. A qualification tester should test the cable link (or channel) to prove it meets transmission requirements based on one of the IEEE 802.3 (Ethernet) Standards, such as 100Base-Tx or 1000Base-T.

A qualification test tool may be designed to test only a copper link or only a fiber link, or the test tool may be designed to provide testing over both media types. Qualification test tools can be used successfully to evaluate and troubleshoot network performance or problems when it is known that the original cable link had been certified before the active equipment was attached. Qualification testing verifies that the cable can support the data rate (measured in bits per second) required by the connected electronics. **See Figure 10-5.**

VERIFICATION FIELD TESTING

Verification field testing assures that the cabling link meets basic connection or continuity requirements. Verification measurements include proper wire pair-to-pair continuity (wire mapping), testing for shorted or open conductors, cable length, power over Ethernet (PoE), etc.

Several verification tests may be performed by a single test tool, or the test tool might perform just one or two of these tests. It is important to understand that verification testing does not provide measurements of the transmission quality of the installed cable link, nor does a verification tool provide information on the bandwidth performance of the installed cable link.

Qualification and verification test tools are typically used for troubleshooting cabling systems. They are much less expensive than certification test sets and are more affordable to have in a tool bag or service vehicle for daily use. **See Figure 10-6.**

Figure 10-5 SignalTEK NT

Figure 10-5. Qualification testers provide proof of transmission performance as a pass/fail per the standards. *Courtesy of IDEAL Networks*

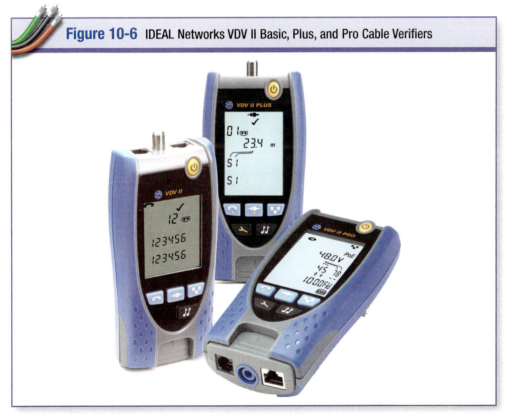

Figure 10-6 IDEAL Networks VDV II Basic, Plus, and Pro Cable Verifiers

Figure 10-6. Verification testers may test for wire map, TDR length/distance to fault, and network/PoE detection. *Courtesy of IDEAL Networks*

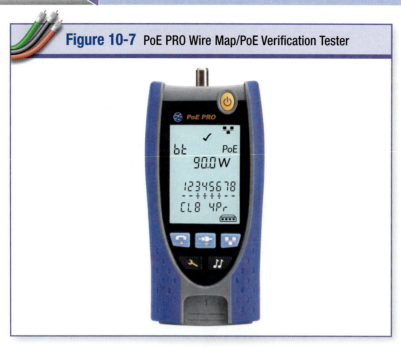

Figure 10-7 PoE PRO Wire Map/PoE Verification Tester

Figure 10-7. PoE verification testers typically have verification features with essential PoE testing: standard (802.3 af/at/bt) along with pass/fail for available power. *Courtesy of IDEAL Networks*

TIP

To learn more about verification testing and the proper use of the testing tools, see the *electrical training ALLIANCE's* course on Test Instruments. This course uses the Test Instruments Textbook (S471) or the Test Instruments and Applications Second Edition text (S571) published by American Technical Publishers (ATP).

VARIOUS TYPES OF CABLING VERIFIERS

There are a multitude of testers designed to check the cabling system throughout the installation and during troubleshooting. Some testers are made to verify the power received at the powered device in a PoE installation. **See Figure 10-7.** Testers may have available accessories to make the job easier. **See Figure 10-8.** Other testers will provide a view of what a security camera sees. **See Figure 10-9.** Others may help locate a mislabeled cable.

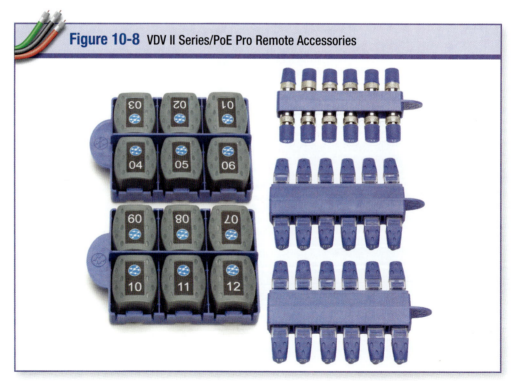

Figure 10-8 VDV II Series/PoE Pro Remote Accessories

Figure 10-8. Kits with RJ45 remotes/terminators and coax remotes allow technicians to test more efficiently. *Courtesy of IDEAL Networks*

Figure 10-9 SecuriTEST IP Camera Installation Tester

Figure 10-9. Camera installation testers may have the ability to power up, configure, monitor, troubleshoot, and document security camera installations. Courtesy of IDEAL Networks

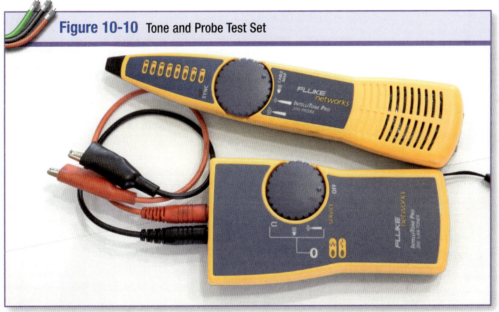

Figure 10-10 Tone and Probe Test Set

Figure 10-10. Tone and probe test sets come in a variety of styles with differing accessories and abilities.

A tone and probe set contains a tone generator and an inductive amplifier. Most "toners" generate a continuous and/or a warbling tone. The toner places one of these signal tones onto the wire, where it is detected by the inductive pickup of the probe. The probe has a built-in speaker (some may also have an earphone jack) allowing the technician to hear the tone as the inductive probe nears the target wire. **See Figure 10-10.**

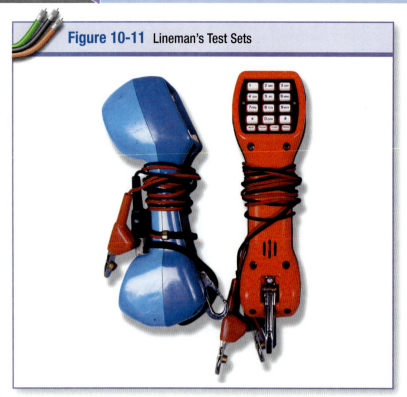

Figure 10-11 Lineman's Test Sets

Figure 10-11. Lineman's test sets were often called "butt sets" because the technician could clamp onto a line and "butt into" an active conversation.

A lineman's test set allows the technician to access and test POTS circuits (local loop telephone lines). Some test sets are considered "data safe" and can safely draw dial tone from a DSL or test an ISDN/BRI circuit without bringing the circuit down. Do not use a regular test set on any data circuit. With the test set, the technician can monitor an active call, answer incoming calls, and/or make outgoing calls. **See Figure 10-11.**

STANDARDS-COMPLIANT CABLING TRANSMISSION AND TEST REQUIREMENTS

Standards-based field testing means that the contractor, the installer, and the owner all know what the outcome of field testing will be when the system has been installed properly. This also means that no matter whose tester was used for the certification test, the outcomes will be the same. It is very important for the contractor and the installer to know what is required in the standards-based test and what standards for testing have been called for in the job specifications.

Not only will the testing prove cabling performance, it will also assist the installer should any troubleshooting be required.

Certifying the Installed Cabling System

The TIA-568.2-D Standard specifies testing parameters for all category cables. Field certification testing will verify cable transmission performance based on the cable's characteristics, connecting hardware, patch cords, and cross-connections. Performance also depends on the quality of workmanship provided during installation of the cable through the pathways and spaces and the quality of the terminations.

There are two main reasons for certification testing. One is that the cabling system is not a complete end-to-end system (final product) purchased by the end user or customer. It is an on-the-job assemblage of cable, patch panels, telecommunications outlets, patch cords, etc. This is an installed system, which becomes the customer's final product. Therefore, the customer needs assurances that the final product has been installed properly. The second reason for certification is to assure the manufacturer of the cable and cabling hardware that the system has been installed properly before the company issues a cabling system warranty to the customer.

Basically, certification testing assures that the cabling system will not be the cause of any network problems and it provides a guarantee that the cabling system will support the network specifications.

Test Configurations

Three test configurations are identified within TIA-568.2-D; they are the *channel test configuration*, the *permanent link test configuration*, and the *modular plug terminated link (MPTL) configuration*.

Channel Link

The channel test configuration is to be used by system designers and users of data communications systems to verify the performance of the overall channel. The channel includes up to 90 meters (295 ft) of horizontal cable, a work area equipment cord, a telecommunications outlet/connector, an optional transition/consolidation point, and two connections in the telecommunications room for a total of 100 meters (328 ft).

To perform the channel test, first make sure that the certification tester has its channel adapter in place on both the certifier and the remote. Calibrate the testers, using the manufacturer's recommended procedures. Plug the patch cord that will be used at the work area outlet into the certifier remote and the patch cord used for the equipment end of the channel into the certifier itself. Run the test according to the certifier manufacturer's instructions. When the test has been completed, record the results and leave all of the patch and equipment cords that were used as part of the channel under test in place. **See Figure 10-12.**

Permanent Link

The permanent link test configuration is to be used by installers and users of data telecommunications systems to verify the performance of permanently-installed cabling. The permanent link consists of up to 90 meters (295 ft) of horizontal cabling and one connection at each end. It may also include an optional transition or consolidation point connection. The permanent link test excludes the cable portion of the test instruments on both the certifier and remote.

To perform the permanent link test, first make sure that the certification tester has its permanent link adapter in place on both the certifier and the remote. Calibrate the test set using the manufacturer's recommended procedures. Plug the link adapter of the remote into the work area outlet and the link adapter of the certifier into the appropriate outlet in the telecommunications room at the patch panel or cross-connect. (Note: If cross-connections are used, then channel testing is recommended.) Run the permanent link test according to the certifier manufacturer's instructions. When the test has been completed, record the results. **See Figure 10-13.**

Modular Plug Terminated Link

A modular plug terminated link (MPTL) configuration is a cabling topology where the far-end connector is a field-installed plug (male) instead of a typical jack (fe-

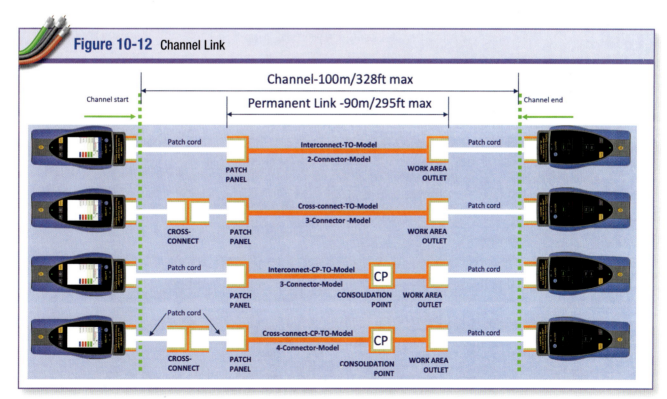

Figure 10-12. *The channel test configuration includes a test of the patch and work-station cables to prove the entire channel works properly. Courtesy of IDEAL Networks*

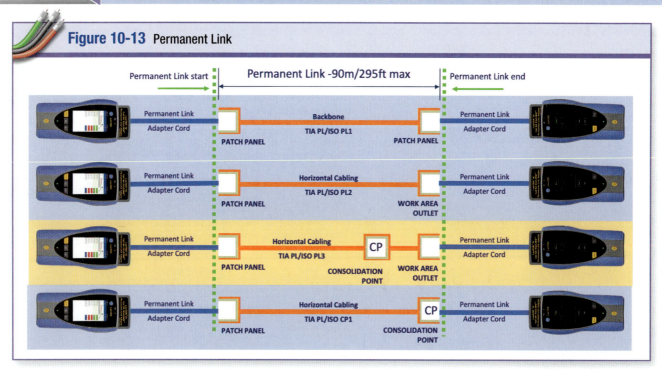

Figure 10-13 Permanent Link

Figure 10-13. The permanent link test configuration tests only the permanently-installed horizontal cabling from the work area outlet to the termination at the patch panel. *Courtesy of IDEAL Networks*

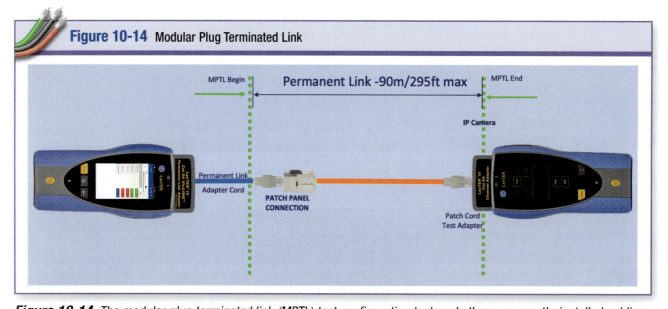

Figure 10-14 Modular Plug Terminated Link

Figure 10-14. The modular plug terminated link (MPTL) test configuration tests only the permanently-installed cabling from the modular jack on one end to the field-installed plug (RJ45) on the other end. *Courtesy of IDEAL Networks*

male). MPTLs are used when the cabling is plugged directly into fixed equipment such as wireless access points, CCTV cameras, access control devices, or other devices where installing a typical outlet is not practical. When certifying MPTLs, the test adapter that the plug end of the link connects to must be a special patch cord test adapter, not the typical channel adapter.

Patch cord test adapters use jacks that are specifically designed to test the performance of the connected plug. Channel adapters do not accurately measure the performance of a connected plug. Run the MPTL test according to the manufacturer's instructions. When the test has been completed, record the results. **See Figure 10-14.**

WHAT IS MEASURED DURING THE TEST

Copper certification performs thousands of complex electrical measurements against industry parameters, standards, defined frequency, and accuracy levels. This certification process proves each cable meets the minimum performance of its category rating. The whole idea of certification testing is to make sure that the transmitted signal can get to the receiver at a higher level than any noise that may also be on the cable. This is a measure of signal strength over various specified frequencies. While this may be an overly simplified explanation, the basic premise is to have more signal than noise after the cabling system has been installed.

The following is a list of the field test parameters, either measured or calculated by the certification tool, used to check whether the signal level is within specification or if the noise floor is too high.

The primary field test parameters are:

- Wire map
- Length
- Insertion loss
- Pair-to-pair near-end crosstalk (NEXT)
- Power sum near-end crosstalk (PSNEXT)
- Attenuation-to-crosstalk ratio (ACR)
- Attenuation-to-crosstalk ratio, far-end (ACRF)
- Power sum attenuation-to-crosstalk ratio, far-end (PSACRF)
- Return loss
- Propagation delay
- Delay skew
- DC loop resistance

Insertion loss, near-end crosstalk, attenuation-to-crosstalk ratios, and return loss are all derived from swept/stepped frequency measurements or equivalent measurement methods.

> ### REMINDER
>
> For a more in-depth explanation of the field test parameters, please review Chapter 4, "Structured Cabling System Performance."

The certification tester physically measures certain parameters and then calculates the rest based on these measurements. It is important to understand that the measured test parameters need to be taken from both ends of the link or channel during the test. The measured test parameters are performed from the certifier and then also from the remote, giving a bidirectional test.

Wire Map

The wire map test is intended to verify pin-to-pin termination at each end. The wire map test indicates:

- Continuity to the remote end
- Shorts between any two or more conductors
- Reversed pairs
- Split pairs
- Transposed pairs
- Any other mis-wiring

Length

There are two ways to measure the length of the cable under test: physical length and electrical length. Physical length is the sum of all of the cables between two end points. This can be calculated by looking at the measurement markings printed on the cable. The electrical length is derived from the propagation delay of signals and depends on the construction and material properties of the cable.

The pass or fail criteria is based on the maximum length allowed for the channel or permanent link plus the nominal velocity of propagation (NVP) uncertainty of less than or equal to 10%. Calibration of NVP is critical to the accuracy of length measurements.

Insertion Loss

Insertion loss is a measure of signal loss in the permanent link or channel. Worst-case insertion loss relative to the maximum insertion loss allowed shall be reported.

The channel insertion loss is derived as the sum of:

- insertion loss of four connectors
- insertion loss of 10 meters (33 ft) of patch, work area, and equipment cords at 20°C

- insertion loss of 90 meters (295 ft) cable segment at 20°C

The permanent link insertion loss is derived as the sum of:

- insertion loss of up to three connectors
- insertion loss of 90 meters (295 ft) cable segment at 20°C

Pair-to-Pair Near-End Crosstalk (NEXT)

Pair-to-pair NEXT is a measure of signal coupling from one pair to another within a 100-ohm twisted-pair cabling link and is derived from swept/stepped frequency or equivalent voltage measurements.

Power Sum Near-End Crosstalk (PSNEXT)

PSNEXT takes into account the combined crosstalk (statistical) on a receive pair from all near-end disturbers operating simultaneously. (In testing, higher numbers are better.)

Pair-to-Pair Attenuation-to-Crosstalk Ratio (ACR) and Far-End Crosstalk (FEXT) Parameters

FEXT is a measure of the unwanted signal coupling from a transmitter at the near-end into another pair measured at the far-end. This measurement is not displayed on the tester but is used to calculate other measurements.

Pair-to-pair ACR is expressed in decibels as the difference between the measured FEXT loss and the insertion loss of the disturbed pair.

Power Sum Attenuation-to-Crosstalk Ratio, Far-End (PSACRF)

PSACRF is the undesired coupling of signal energy from a number of simultaneously-transmitting pairs into a receiving pair. PSACRF is calculated at the opposite end of the cable from where the transmitter is located. There are four PSACRF results for each cable end.

This is an especially important measurement for systems that utilize more than one pair for transmitting signals, while using multiple adjacent pairs, or the same pairs, for receiving signals. PSACRF is the sum of the attenuation-to-crosstalk ratio, far-end (ACRF) power from all other conductor pairs in the cable. (In testing, higher numbers are better.)

Return Loss (RL)

Return loss is a measure of the reflected energy caused by impedance variations in the cabling system. Return loss is a measurement of the amount of signal energy that is reflected back to a transmitter as it travels down a twisted pair while encountering impedance discontinuities.

The measurement is expressed in decibels and is a ratio of the amplitude of the transmitted signal with respect to the amplitude of the reflected signal. Impedance discontinuities occur at connections where cable is terminated to plugs or jacks and within the plug/jack connection. An impedance discontinuity can also occur if a cable has exceeded its bend radius, is kinked, or is otherwise damaged. When a transmitted signal pulse hits one of these structural discontinuities, return loss occurs.

Too much RL noise added to the PSNEXT and PSACRF adds to the noise, resulting in increased bit error rate, lower signal-to-noise ratio, less network operating margin, and more downtime. (In testing, a lower absolute value is better.)

Propagation Delay

Propagation delay is the time it takes for a signal to propagate from one end of the cable to the other. Delay is measured in nanoseconds.

Because each of the four pairs in a 4-pair cable has a different number of twists, the electrical length of each pair is different. The pair with the longest (loosest) twist length is electrically shorter than the pair with the shortest (tightest) twist length. Therefore, the amount of time it takes for a signal to travel from the transmitting end to the receiving end of a pair (delay) is different for each of the pairs.

Delay Skew

Delay skew is the difference between the amount of delay for the fastest (shortest twist lengths) and slowest (longest twist lengths) pairs within a cable.

DC Loop Resistance Unbalance

DC loop resistance is the total resistance through two conductors that are looped at one end of the length, measured within each pair and between pairs. The resistance will vary with the conductor's length and is based on the conductor's diameter. This is an important measurement when power is added to the signal. For example, 10/100Base-T transmission only uses two pairs of a cable for data while the other two are used for power (PoE), so unbalance is not an issue.

Faster transmission, such as Gigabit Ethernet, requires the use of all four pairs for bi-directional transmission. When the application also requires power, that power must be transmitted on those same pairs. PoE typically does not cause a problem with data transmission; however, too much DC resistance unbalance (DCRU) in a connection can distort data signals, causing re-transmits and loss of transmission. With excessive DCRU, PoE may also stop functioning.

10 GIGABIT ETHERNET

The most-used media options for 10 Gigabit Ethernet deployment today include:

- 10GBase-T over category 6 cabling is a twisted-pair solution but only to 55 meters
  - Category 6 cable is specified to 250 MHz at 100 meters, but due to added issues with crosstalk, support of 10G is limited
- 10GBase-T over category 6A cabling is a twisted-pair solution for up to 100 meters
  - Category 6A cables are made with larger conductors to accomplish transmission over a longer distance
  - Category 6A cable is specified to 500 MHz at 100 meters
- 10GBase-SR over 50-micrometer multi-mode (OM3) fiber at 850 nanometers using a vertical cavity surface emitting laser (VCSEL) transmitter to 300 meters
- 10GBase-SR over 50-micrometer multi-mode (OM4) fiber at 850 nanometers using a VCSEL transmitter to 400 meters
- 10GBase-LR over single-mode fiber (OS2) at 1,310 nanometers with a laser transmitter up to 10 kilometers
- 10GBase-ER over single-mode fiber (OS2) at 1,550 nanometers with a laser transmitter up to 40 kilometers

Fiber is always an excellent choice when practical, especially for longer distances.

Considerations for 10GBase-T

There are several issues that must be taken into consideration when deploying a 10GBase-T network. 10GBase-T requires a more sophisticated encoding process (PAM-16) that allows for less tolerance in signal levels, and because it is very noise-sensitive, it requires the use of higher operational frequencies. This induces two distinct problems:

- Higher frequencies increase insertion loss (attenuation) and limit the channel distance as well as reduce the attenuation-to-crosstalk ratio.
- Higher frequencies equal higher crosstalk, to the point where a previously unimportant phenomenon becomes a critical issue.

All test parameters for 10GBase-T must be tested in the field to 500 MHz. All types of crosstalk are exacerbated as frequencies increase. Attenuation (insertion loss) is also affected by increased frequency, leading to problems on long-length channels. Larger diameter conductors are necessary to lower signal attenuation on lengths up to 100 meters.

During the development of cabling standards to support 10 Gigabit Ethernet, alien crosstalk (AXT) was discovered to be a primary source of interference. AXT is the coupling of unwanted signals between cables in a bundle and is most prevalent at frequencies above 400 MHz. **See Figure 10-15.** Running 10 Gigabit Ethernet over category 6 or category 6A cabling would have required AXT field testing to ensure compliance. However, cable manufacturers improved the design of category 6A cable to effectively eliminate AXT concerns. The primary design element is using thicker cable jackets to increase the distance between

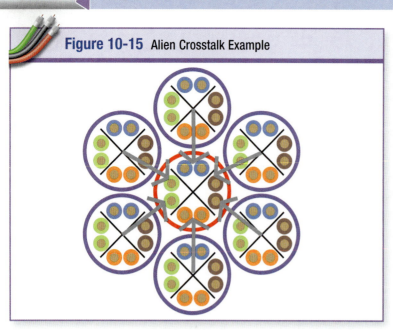

Figure 10-15 Alien Crosstalk Example

Figure 10-15. *In the event of of alien crosstalk, multiple disturber cables may affect a single cable in a bundle. Courtesy of IDEAL Networks*

conductors in cable bundles. The radiated energy is weak enough that even two or three millimeters of additional spacing between signal conductors will eliminate AXT concerns.

40 Gigabit Ethernet Over Copper Cabling

To support 40 Gigabit Ethernet applications on twisted-pair cabling, both the ANSI/TIA and ISO/IEC standards bodies developed specifications for a new type of cabling that would operate at up to 2,000 MHz (2 GHz). In November 2017, ISO/IEC updated the 11801-1 standard to add Class I and Class II cabling, and in July 2016, ANSI/TIA released the ANSI/TIA-568-C.2-1 addendum to add category 8 cabling.

A significant difference between category 8, Class I and II, and the previous category cabling is the supported distance. All previous systems allow permanent links of up to 90 meters (295 ft), while category 8 is limited to just 30 meters (98 ft). The reason for the reduced length is that insertion loss at the frequencies required to support 40 Gigabit Ethernet is so high that the signal-to-noise ratio (SNR) at distances greater than 30 meters is unusable.

Category 8 is primarily intended for use in data centers where 30 meters is long enough for most links between equipment cabinets. (Category 8 cabling can be installed as 90-meter permanent links, though the bandwidth is limited to 10 Gigabit Ethernet applications.)

SUMMARY

The testing of a structured cabling system ensures the customer that the final product will perform to the level specified within the construction documents.

Occasionally, a basic verification test will be acceptable to the customer; however, a certification test ensures the system will perform to the minimum standards set by ANSI/TIA-568.2-D or ISO/IEC 11801-1. It also ensures the cabling will not be the cause of any network failures, either now or in the future.

Using a recently calibrated test set and correctly choosing the test configuration for the cable category and cable type at initial set-up are essential to receiving accurate test results. A digital or printed copy of the certification test results is required for warranty coverage by the cable and hardware manufacturer.

REVIEW QUESTIONS

1. Which industry standard defines the testing parameters for certification testing?
   a. TIA-568.0-D
   b. TIA-568.2-D
   c. TIA 569-E
   d. TIA-607-C

2. Which test level is required for a manufacturer warranty?
   a. Certification
   b. MPTL
   c. Qualification
   d. Verification

3. Which test configuration includes the patch and work-station cables?
   a. Certification
   b. Channel link
   c. Modular plug terminated link
   d. Permanent link

4. What is the test parameter that measures the difference between the measured FEXT loss and the insertion loss of the disturbed pair?
   a. Attenuation-to-crosstalk ratio
   b. Delay skew
   c. Power sum near-end crosstalk
   d. Return loss

5. What type of tester will perform a load test to determine if the power available at the device end of the cable meets the minimum requirement with a pass/fail indication?
   a. Certification test set
   b. Channel link
   c. Modular plug terminated link
   d. Power over Ethernet tester

6. What type of tester would be useful in locating a mis-labeled cable?
   a. Certification test set
   b. Qualification test set
   c. Toner and probe test set
   d. Verification test set

7. All testing parameters are important; however, when power is added to the signal, __?__ becomes even more important.
   a. DC loop resistance unbalance
   b. near-end crosstalk
   c. propagation delay
   d. return loss

8. What test parameter is a measure of reflected signal because of impedance variations?
   a. DC loop resistance unbalance
   b. Near-end crosstalk
   c. Propagation delay
   d. Return loss

9. Category 6 cable must be field tested to __?__, while category 6A must be tested to __?__.
   a. 100 MHz / 200 MHz
   b. 200 MHz / 500 MHz
   c. 250 MHz / 500 MHz
   d. 500 MHz / 1,000 MHz

10. __?__ is the coupling of unwanted signals between cables in a bundle and is most prevalent at frequencies above 400 MHz.
   a. AXT
   b. FEXT
   c. NEXT
   d. PSNEXT

Structured Cabling Systems Applications

Introduction

There are several different structured cabling applications used in a commercial building. Installers should be familiar with the layout for wireless, Power over Ethernet (PoE), telecommunications enclosures, and open office architecture along with a few others.

Appendix

Table of Contents

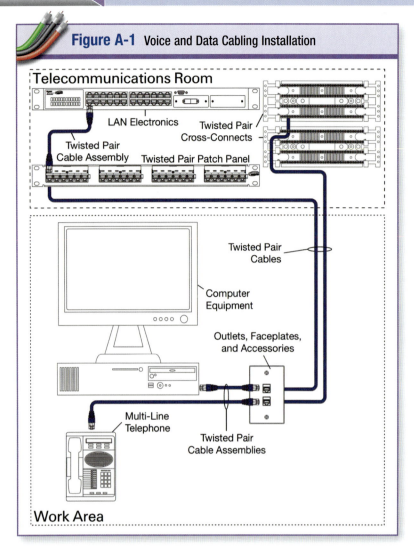

Figure A-1. *A typical voice and data cabling installation runs from the telecommunications room (TR) to the work area (WA) using UTP cabling.*
Courtesy of Tyco Electronics

VOICE AND DATA APPLICATION

Telecommunications rooms, at a minimum, supply voice and data services to the work area. **See Figure A-1.** The following components make up a basic voice and data structured cabling system:

- LAN electronics (switches and/or routers typical)
- Twisted-pair patch panels
  - May be 24 ports or higher
  - May be rated for category 5e, 6 or 6A
- Twisted-pair cross-connects
- Twisted-pair cable assemblies (patch and work-station cables)
  - May be rated category 5e, 6 or 6A
- Twisted-pair cables
  - May be rated category 5e, 6 or 6A
- Outlets, faceplates, and accessories
- Multi-line telephone
- Computer equipment

TELECOMMUNICATIONS ENCLOSURES

A telecommunications enclosure (TE) is essentially a small TR. It is not meant to replace a TR but to be used in conjunction with a TR when distance or other limitations are a consideration. **See Figure A-2.**

- Backbone optical fiber cabling
- Twisted-pair/optical fiber consolidation points and accessories

Voice Applications

In order to connect a telephone set to a switch using a structured cabling system, a 2-, 4-, 6-, or 8-wire cord is used to connect the phone set to the telecommunications outlet at the work area. A single-line analog phone utilizes only one pair of wires for operation. When plugged into an 8-position/8-conductor (8P8C) modular jack, the plug on the end of the cord will center itself in the jack in such a way that the center two pins of the jack (4 and 5) are accessed. These two pins are connected to pair 1 of the UTP horizontal cable. At the TR, this pair is cross-connected to an assigned pair in the voice backbone cable, using paired cross-connect wire. At the voice equipment room, the backbone pair is cross-connected to the assigned switch port on the equipment terminal block.

Two-line analog phones use 4-, 6-, or 8-wire cords. Line 1 is served by pins 4 and 5 of an 8P8C modular jack, while Line 2 is served by pins 3 and 6. Each line requires a separate twisted pair in the horizontal cable. At the telecommunications room, the circuit on pins 4 and 5 of the jack appears as pair 1. The circuit on pins 3 and 6 of the jack appears as pair 2, if the jack is wired per the T568A pinout, or as pair 3 if the jack is wired per the T568B pinout. Two 1-pair cross-connections are required (one for each line) at the TR and at the ER.

- Twisted-pair/optical fiber patch cable assemblies
  - Category 5e, 6, or 6A patch cable assemblies
  - SC or LC duplex patch cable assembly, 50/125 micrometer multimode fiber
- Optical fiber rack mount enclosures
- LAN electronics (switches and/or routers typical)
- Twisted-pair cables
  - May be rated category 5e, 6 or 6A
- Outlets, faceplates, and accessories
- Computer equipment

Additions to the TIA-568 and TIA-569 Standards have allowed a type of network architecture called a telecommunications enclosure (TE). This architecture allows an effective combination of optical fiber for longer distances and twisted-pair cables for work area outlets. While the concept was initially designed to simplify modular office moves, adds, and changes, the concept offers advantages in other types of installations.

The base of the telecommunications enclosure architecture is the TE itself. Optical fiber backbone cabling is run from the equipment room, through the telecommunications room, and into the TE where it is terminated and patched to a switch. The output of the switch (RJ45 ports) is then patched to the horizontal twisted-pair cable to the outlet in the work area.

At first glance, this appears to be a displaced TR. However, the requirements for the TE are not the same as for the TR, so there is some additional flexibility when using the TE. It is important to note that the standards do not allow a network without a TR, but in some installations where multiple TRs are needed, TEs may be used in place of additional TRs.

A TE should support only an area of 335 meters squared (3,600 ft²) and meet the additional requirements of TIA-569-E.

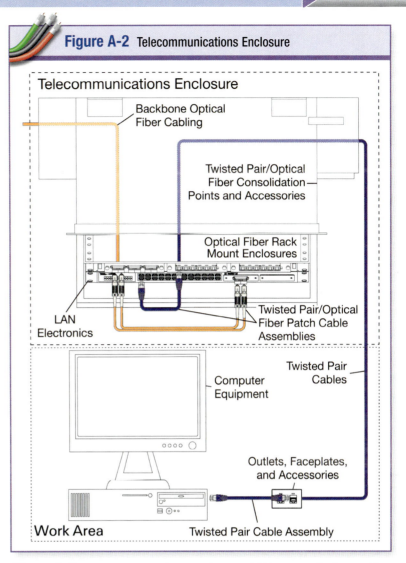

Figure A-2 Telecommunications Enclosure

Figure A-2. Telecommunications enclosures (TEs) are used when cable distance limitations are a concern or in areas with a lot of cabling activity. Courtesy of Tyco Electronics

F/UTP APPLICATION

Shielded product solutions offer higher electrical performance and virtually eliminate alien crosstalk (AXT) associated with 10 Gigabit Ethernet transmission. This system layout meets all ANSI/TIA category 6A performance standards and IEEE 802.3an 10 Gigabit Ethernet requirements. **See Figure A-3.**

- Twisted-pair cables
  - Category 6A F/UTP (or similar) cable

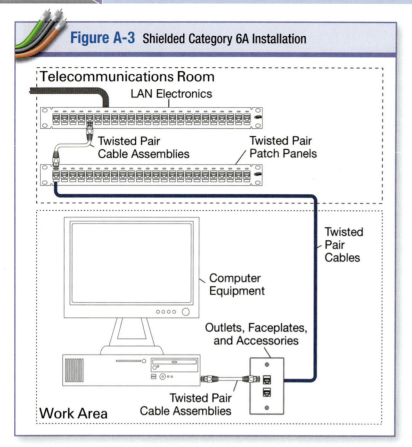

Figure A-3 Shielded Category 6A Installation

Figure A-3. An F/UTP installation should include all shielded components. Courtesy of Tyco Electronics

- Twisted-pair patch panels
  - Category 6A shielded patch panel, 24-port or larger
- Twisted-pair cable assemblies
  - Category 6A shielded patch cable assembly
  - Category 6A F/UTP cable
- Outlets, faceplates, and accessories
- Computer equipment

OPEN OFFICE APPLICATION USING A CONSOLIDATION POINT

Using a consolidation point with an open office can save on time and cabling when rearranging office furniture. **See Figure A-4.**

- Twisted-pair patch panels
  - May be rated category 5e, 6 or 6A
- Optical fiber rack mount enclosures and connectors, couplers, and adapters

- Optical fiber cabling
  - Horizontal cable, 2-fiber (dual), 50/125 micrometer multimode, OFNP
- Twisted-pair cables
  - May be rated category 5e, 6 or 6A
- Twisted-pair consolidation points and accessories
  - Modular consolidation point UTP or optical fiber
- Twisted-pair patch cable assemblies
  - May be rated category 5e, 6 or 6A
- Optical fiber cable assemblies
- Outlets, faceplates, and accessories

The consolidation point functions as an interconnect between two sections of the horizontal cabling: the main cabling from the telecommunications closet and shorter distribution cables which serve outlets in the work area. The consolidation point enclosure is typically mounted in the ceiling or under access flooring. When furniture moves are required, only the shorter cables must be reconfigured. Consolidation point modules are available for both optical fiber and twisted-pair cabling.

ANSI/TIA-568.1-D, *Commercial Building Telecommunications Cabling Standard,* Part 1: General Requirements defines two styles of horizontal cabling systems specifically geared towards modular furniture: the consolidation point (as discussed before) and the multi-user telecommunications outlet assembly, or MUTOA. Both of these cabling system practices are designed to lessen the amount of "re-cabling" involved in modular furniture moves.

OPEN OFFICE APPLICATION USING A MULTI-USER TELECOMMUNICATIONS OUTLET ASSEMBLY (MUTOA)

A MUTOA is a connection point to the LAN that is made accessible to users for use instead of an individual work area location. **See Figure A-5.**

- Twisted-pair patch panels
  - Category 5e or better

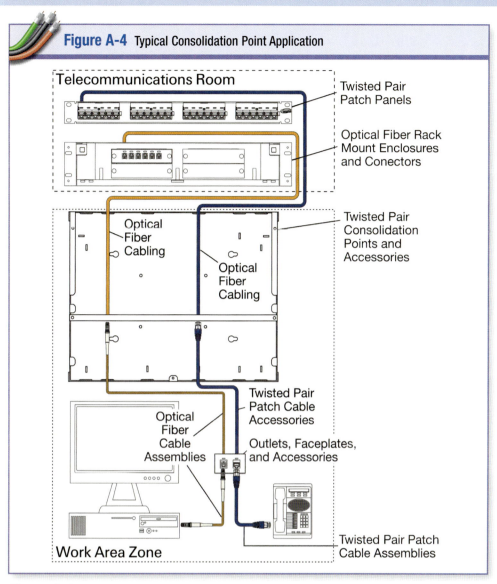

Figure A-4 Typical Consolidation Point Application

Figure A-4. Consolidation points in open office applications can be useful for areas with a lot of moves, adds, and changes work. *Courtesy of Tyco Electronics*

- Optical fiber rack mount enclosures
- Optical fiber cabling
  - Horizontal cable, 2-fiber (dual), 50/125 micrometer multi-mode, OFNP
- Outlets, faceplates and accessories
  - Multi-port surface-mount type boxes are common
- Twisted-pair cables
  - Category 5e, 6, or 6A
- Optical fiber cable assemblies
- Twisted-pair cable assemblies
  - Category 5e patch cable assemblies or better

The MUTOA is meant to be accessible to users and replaces individual work area outlets. Multiple horizontal cables terminate at an outlet where the individual users connect to the LAN with work area patch cables. TIA/EIA-568-B.1 allows for work area patch cords longer than the three meters specified elsewhere in the Standard to accommodate multiple users. Each outlet should be labeled with the longest allowable patch cord. **See Figure A-6.**

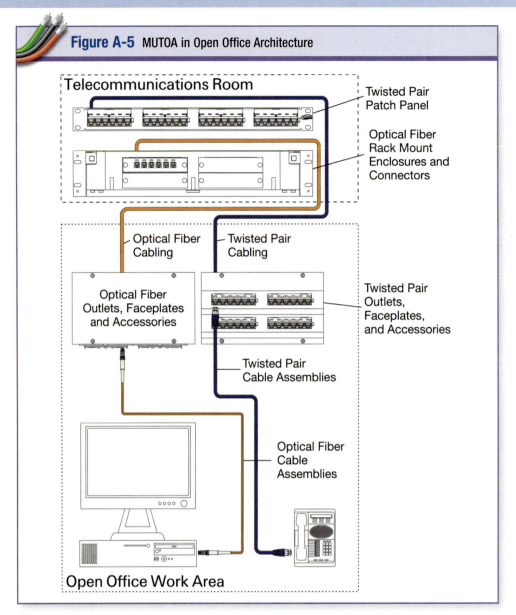

Figure A-5 MUTOA in Open Office Architecture

Figure A-5. *A MUTOA may be used in an open office architecture.* Courtesy of Tyco Electronics

Figure A-6 Work Area Cables Based on Length of Permanent Link

Length of Horizontal Cable m (ft)	Maximum Length of Work Area Cable m (ft)	Maximum Combined Length of Work Area Cables, Patch Cords and Equipment Cables m (ft)
90 (295)	3 (10)	10 (33)
85 (279)	7 (23)	14 (46)
80 (262)	11 (36)	18 (59)
75 (246)	15 (49)	22 (72)
70 (230)	20 (66)	27 (89)

Figure A-6. *Work area cable length may be extended if the permanent link length is less than 90 meters (295 ft).*

WIRELESS APPLICATION

The wireless LAN is quite similar to a wired LAN with the addition of wireless access points for mobility. **See Figure A-7.**

- LAN electronics (SNMP managed ethernet switch)
- Twisted-pair patch panels
  - May be rated category 5e, 6 or 6A
- Outlets, faceplates, and accessories (MPTL optional)
- Wireless access point(s)
- Twisted-pair patch cable assemblies
  - May be rated category 5e, 6 or 6A
- Twisted-pair cables
  - May be rated category 5e, 6 or 6A

Mobility within the work environment has become a necessity in today's workplace. Whether moving employees, adding additional workspace, or in the event of time constraints when the need for new cable installation arises, high-speed wireless networking may be the answer. The indoor wireless system provides high-speed network connectivity without cabling. Functioning in the same manner as a wired LAN, the wireless LAN can be utilized by itself or as a complement to an existing wired LAN. The network remains safe, secure, and reliable.

POWER OVER ETHERNET (PoE) OVER TWISTED-PAIR CABLING

Power over Ethernet cable is a great way to power devices on the LAN without needing a power outlet installed.

- LAN electronics
  - PoE module
- Twisted-pair cable assemblies
  - Category 5e or better patch cable assemblies
- LAN electronics
- Twisted-pair patch panels
  - Category 5e or better
- Twisted-pair cables
  - Category 5e or better
- Outlets, faceplates, and accessories
- PoE ready VoIP phone
- PoE ready wireless access point (WAP)

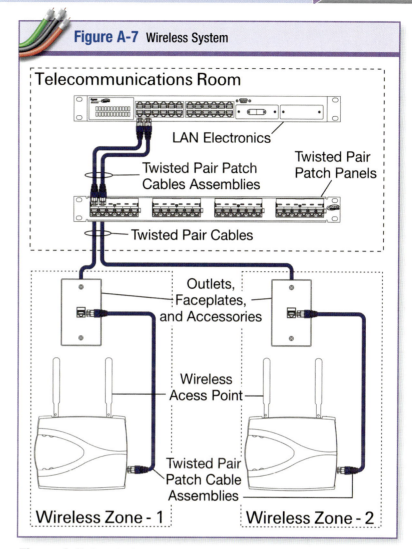

Figure A-7 Wireless System

Telecommunications Room

LAN Electronics

Twisted Pair Patch Panels

Twisted Pair Patch Cables Assemblies

Twisted Pair Cables

Outlets, Faceplates, and Accessories

Wireless Acess Point

Twisted Pair Patch Cable Assemblies

Wireless Zone - 1   Wireless Zone - 2

Figure A-7. *A typical wireless system allows mobility in the workplace.* Courtesy of Tyco Electronics

Consider a system with a standard LAN switch connected to an 8-port PoE midspan device which adds power to the data signal before connection to the patch panel. **See Figure A-8.** Category 5e or better cabling is installed to the outlet where patch cords connect to standard work area devices or IEEE 802.3af-compliant devices, such as a wireless access point.

A PoE midspan device offers a cost-effective solution for power to remote devices like VoIP phones, WLAN access points, security devices, etc. The PoE midspan device injects DC power into the cable, thus eliminating the cost for powered switch ports, and it allows routing power only to the outlets where power is needed.

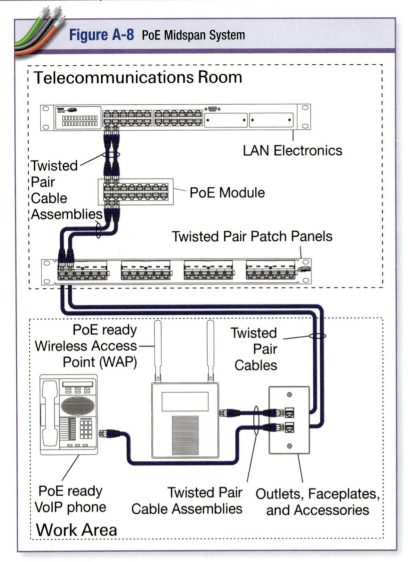

Figure A-8. *A typical Power over Ethernet system may utilize a midspan device.* Courtesy of Tyco Electronics

A PoE midspan device offers some of the following advantages over end-span devices like powered switches:

1. Allows the widest selection of switch vendors and products for PoE—the midspan even adds power to existing switch ports

2. Provides a migration path to PoE without the cost of upgrading to new switches

3. Improves port utilization for PoE ports, and removes the excess heat generation associated with end-span devices

STORAGE AREA NETWORK (SAN)/ DATA CENTER EQUIPMENT APPLICATION

This is a high-density and high-performance application for a copper cabling system.

- Twisted-pair data center/SAN products
  - MRJ21 patch panel, angled, 48-port
- Twisted-pair data center/SAN products
  - MRJ21 cable assembly, 180° back shells
- Twisted-pair patch cable assemblies
  - May be rated category 5e or category 6/6A
- Rack-mounted server

MRJ21 High-Density Copper Cabling System

The MRJ21 connector system solution is a high-density, high-performance, modular copper cabling system. **See Figure A-9.** The matched power sum 24-pair cabling and connector solution provides a variety of applications for up to 12 ports with a single cable. This solution supports any plug-and-play environment, including data centers and zone-cabled or open-office environments. When utilizing a PC board mounted solution, the MRJ21 connector enables much higher switch port density than the common modular plug/jack interface, which can reduce active port costs.

MRJ21 System Attributes

The main attributes of the MRJ21 system are:

- The MRJ21 connector system provides an application-independent copper platform for applications today and into the future, including Gigabit, VoIP, and powered Ethernet.
- The small size of the MRJ21 connector and cabling reduces cable bulk in pathways and spaces, enabling much higher manageable port counts in a smaller space.

- Factory termination and testing of the cassettes and patch panels allows pluggable performance on site; this enables rapid installation or moves, adds, and changes with pre-tested quality.
- The modularity of this solution significantly reduces time to install or migrate from a 10/100Base-T 2-pair platform to a Gigabit Ethernet 4-pair platform. Upgrading is as simple as unplugging one cassette and plugging in the new one. The same cable is re-usable. A 48-port 1U panel can be installed with Gigabit performance in minutes compared to a traditional 48-port discrete RJ45 panel which can take hours to install.
- Cassettes and patch panels break out high-performance 24-pair solutions to the appropriate wiring patterns for standards-based 10/100Base-T, Gigabit Ethernet, and other applications.
- Common cable assembly lengths provide readily-available pluggable solutions, speeding implementation.

CENTRALIZED NETWORK ADMINISTRATION (CNA) SYSTEM

This system saves floorspace and money by having one centralized telecommunications room instead of TRs on each floor of a building.

- Optical fiber switch
- Optical fiber rack-mount enclosures
- Optical fiber cable assemblies
- Optical fiber cable
  - Distribution cable, 24-fiber, 50/125 micrometer multi-mode
  - Horizontal cable, 2-fiber (dual), 50/125 micrometer multi-mode
- Faceplates, outlets, and accessories
- Twisted-pair cross-connects
- Twisted-pair cables
  - Category 5e or better
- Twisted-pair patch cable assemblies
  - Category 5e or better

Centralized network administration (CNA) brings all user connectivity to one centralized telecommunications/equipment room within the building, as opposed to

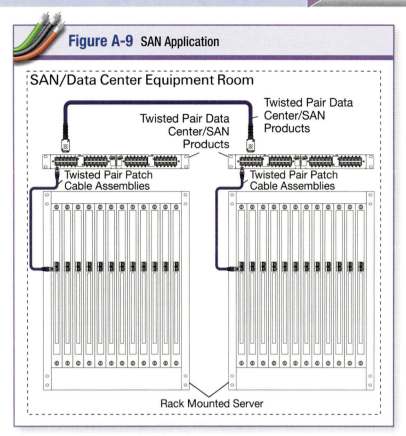

Figure A-9 SAN Application

SAN/Data Center Equipment Room

Twisted Pair Data Center/SAN Products

Twisted Pair Data Center/SAN Products

Twisted Pair Patch Cable Assemblies

Twisted Pair Patch Cable Assemblies

Rack Mounted Server

Figure A-9. The SAN application is a high-density and high-performance application. *Courtesy of Tyco Electronics*

distributed TRs throughout the building. **See Figure A-10.**

The main advantages of using the CNA cabling application are to address the increasing operational costs of network administration and the future requirements of network administration to easily and cost-effectively migrate from the low-speed LANs of today to the high-speed requirements of tomorrow.

FIBER OPTIC CABLING APPLICATIONS CENTRALIZED FIBER NETWORK

In a centralized fiber network, the fiber can route through the TR without termination and straight out to the work area as long as it does not exceed 90 meters (295 ft).

- Optical fiber backbone cable
  - Distribution cable, 24-fiber, 50/125 micrometer multi-mode

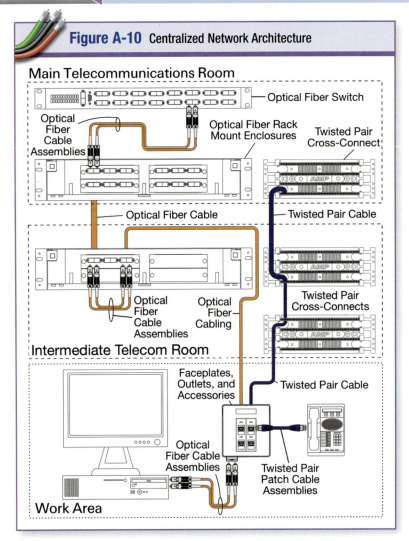

Figure A-10 Centralized Network Architecture

Main Telecommunications Room

- Optical Fiber Switch
- Optical Fiber Cable Assemblies
- Optical Fiber Rack Mount Enclosures
- Twisted Pair Cross-Connect
- Optical Fiber Cable
- Twisted Pair Cable

Intermediate Telecom Room

- Optical Fiber Cable Assemblies
- Optical Fiber Cabling
- Twisted Pair Cross-Connects

Work Area

- Faceplates, Outlets, and Accessories
- Twisted Pair Cable
- Optical Fiber Cable Assemblies
- Twisted Pair Patch Cable Assemblies

Figure A-10. *A typical centralized network architecture eliminates telecommunications room(s) on each floor.* Courtesy of Tyco Electronics

- Optical fiber rack-mount enclosure
- Optical fiber cable assemblies
- Optical fiber LAN equipment with fiber interface
- Optical fiber pull-through cable
  - Horizontal cable, 2-fiber (dual), 50/125 micrometer multi-mode
- Outlets, faceplates, and accessories
- Computer equipment

See Figure A-11. A centralized fiber network is constructed using LAN equipment with a duplex interface, duplex patch cords, plenum fiber cable, and connectors. Note that the pull-through fiber cable is routed through the TR without termination. This is allowed by the TIA-568.0-D Standard as long as the pull-through cable length does not exceed 90 meters.

MPO HIGH–DENSITY OPTICAL FIBER CABLING

MPO connectors and cassettes offer high–density fiber for data centers (DCs) and storage area networks (SANs).

- Rack-mount enclosure and cassettes
- MPO trunk cable assembly, 24-fiber
- Duplex patch cable assembly
- DC/SAN director
- MT-RJ patch cable assembly
- Storage device

Data centers and storage area networks have slightly different conditions than traditional commercial building cabling networks. Unlike commercial building networks, which are distributed networks, the DC/SAN networks are concentrated networks with a premium placed on reliability, density, and speed of installation.

Just as the MRJ21 connector system for copper offers a twisted-pair solution, the MPO products offer a solution for optical fiber networks. These products are based on the high-density, 12-fiber MPO array connector. In the MPO ferrule, 12 fibers are terminated in a connector the same size as a single-fiber SC connector. Thus, the MPO solutions offer the highest-density solution with the added advantages of optical fiber: high data rates, small-diameter flexible cables, and resistance to electromagnetic and radio frequency interference.

Trunk cables are terminated with MPO connectors. Cassettes offer a modular breakout of the fibers in the MPO connector to more common interfaces such as MT-RJ, SC duplex, and LC. To-

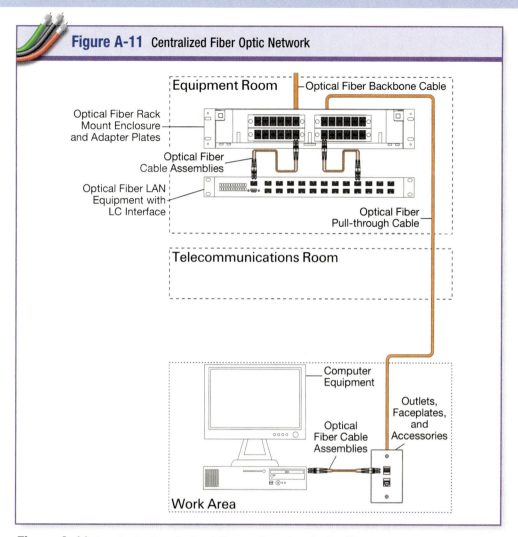

Figure A-11 Centralized Fiber Optic Network

Equipment Room — Optical Fiber Backbone Cable

Optical Fiber Rack Mount Enclosure and Adapter Plates

Optical Fiber Cable Assemblies

Optical Fiber LAN Equipment with LC Interface

Optical Fiber Pull-through Cable

Telecommunications Room

Computer Equipment

Outlets, Faceplates, and Accessories

Optical Fiber Cable Assemblies

Work Area

Figure A-11. *In a typical centralized fiber optic network, the fiber may pass through the TR and terminate at the desk.* Courtesy of Tyco Electronics

gether, they construct a functional link between equipment that can be accomplished rapidly, simply, and without the complexity of a normal move, add, or change.

See Figure A-12. The main distribution area (MDA) is connected to remote equipment in the equipment distribution area (EDA) by a link constructed of common patch cords, a trunk cable, and cassettes in a centralized fiber network. These components can be constructed in many ways to make viable links between equipment, and the interfaces can be mixed and matched on either end of the trunk cable cassettes to match the equipment interfaces. In this example, SC duplex interfaces are used in the MDA, and MT-RJ interface equipment is used in the EDA.

BASEBAND VIDEO OVER UTP WIRING

Baseband video is an un-modulated video signal. This signal does not carry any audio information and uses up the entire media that is carrying the signal (for example, the coax or UTP cable).

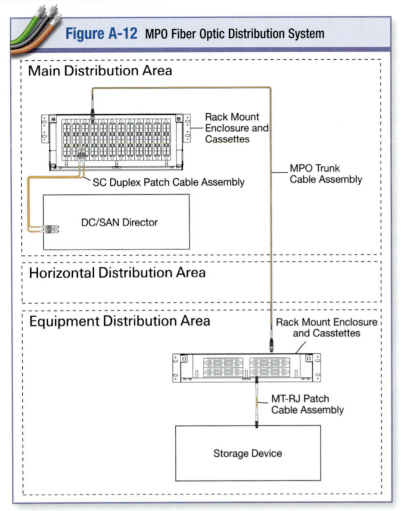

Figure A-12 MPO Fiber Optic Distribution System

Main Distribution Area

Rack Mount Enclosure and Cassettes

MPO Trunk Cable Assembly

SC Duplex Patch Cable Assembly

DC/SAN Director

Horizontal Distribution Area

Equipment Distribution Area

Rack Mount Enclosure and Casstettes

MT-RJ Patch Cable Assembly

Storage Device

Figure A-12. *A typical MPO fiber optic distribution system may be used for high-density applications.* Courtesy of Tyco Electronics

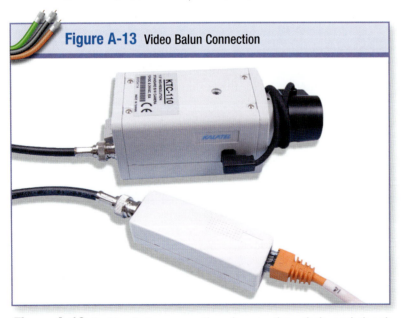

Figure A-13 Video Balun Connection

Figure A-13. *A video balun connection changes the unbalanced signal of the coax to a balanced signal for the twisted-pair cable.*

Baseband video can be transported over the same media as RF (broadband) video but never at the same time. Runs of several hundred feet are possible without amplification, but amplification and distribution of baseband video is different than RF video/audio. Devices utilizing baseband video, such as monitors or CCTV cameras, usually have input or output jacks that are non-threaded RCA style connectors or BNC style connectors. **See Figure A-13.**

Baseband video is also known as *composite video*. Composite video is the analog television signal before a sound signal is added and before both are combined and modulated onto an RF carrier. A composite signal is a composite of three video source signals. They are brightness or luminance, along with the sync pulse, and the color information called chrominance if the image is in color.

To connect a baseband (composite) video source to a video monitor over an unshielded twisted-pair structured cabling system, begin by connecting a short length of coaxial cable between the monitor's BNC or RCA video in jack and the baseband video balun's input. Then, place a modular cord between the balun's modular jack and a modular telecommunications outlet. If stereo audio is also desired, connect a pair of RCA patch cords between the monitor's audio in jacks and the audio jacks on the balun (if the balun is equipped for audio).

At the telecommunications room, cross-connect the horizontal pair that is carrying the video signal to one pair in the backbone cable, using a paired cross-connect wire or a patch cord. If stereo audio is also being carried, an additional two pairs must be cross-connected. At the video source location, a patch cord is used to connect the backbone pair(s) to a video balun. A coaxial cable is used to connect the video balun to the baseband video source (computer, camera, VCR, VTR, etc.). For stereo audio, a pair of RCA patch cords must also be connected between the source's audio out jacks and the audio jacks on the balun. **See Figure A-14.**

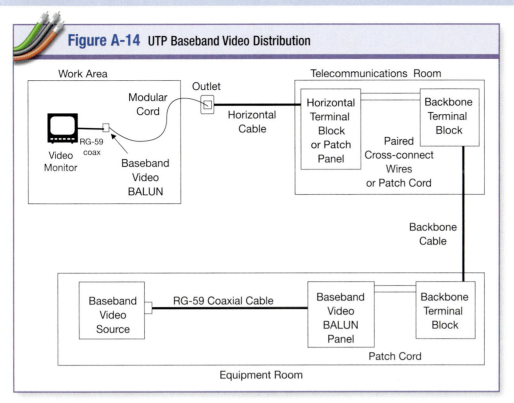

Figure A-14 UTP Baseband Video Distribution

Figure A-14. Baseband video may be distributed over a UTP structured cabling system.

BROADBAND VIDEO OR DATA OVER UTP WIRING

To connect a broadband video signal over an unshielded twisted-pair structured cabling system, begin by connecting the coaxial cable from a broadband balun to the TV or broadband modem's F connector. Next, connect the broadband balun's modular cord to an 8-pin telecommunications outlet. **See Figure A-15.**

At the telecommunications room, cross-connect the horizontal pair used by the balun to one pair in the backbone cable using a patch cord. At the location of the broadband signal pickup, a patch cord is used to connect the backbone pair to the broadband balun. A coaxial cable is used to connect the balun to the broadband signal source. **See Figure A-16.**

Figure A-15 Broadband Video Balun

Figure A-15. A broadband video cable may have a balun on one end and an 8-position/8-conductor connector on the other end.

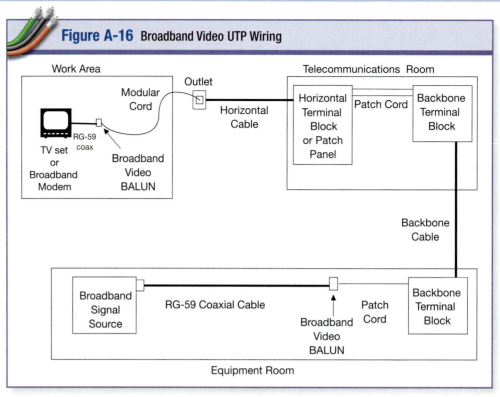

Figure A-16 Broadband Video UTP Wiring

Figure A-16. Broadband video may be transmitted over UTP cabling.

A

Ablative
The development of a hard char that resists the erosion of fire and flames; a characteristic of a firestop when exposed to fire

Adapter
A device that enables any or all of the following:
- Different sizes or types of plugs to mate with one another or to fit into a telecommunications outlet
- The rearrangement of leads
- Large cables with numerous wires to fan out into smaller groups of wires
- Interconnection between cables

Adapter; optical fiber duplex
A mechanical device designed to align and join two duplex optical fiber connectors (plugs) to form an optical duplex connection

Administration
The method for labeling, identification, documentation, and usage needed to implement moves, additions, and changes of the telecommunications infrastructure; includes documentation (labels, records, drawings, reports, and work orders) of cables, termination hardware, patching and cross-connect facilities, pathways, communications rooms, and spaces

Afterset insert
An insert installed after the installation of the concrete floor slab or other flooring material

Alternating current
A current that changes direction at a uniformly repetitious rate

Alternating current equipment ground
A conductor installed from the equipment grounding bus inside the electrical panel to a telecommunications primary bonding busbar or a secondary bonding busbar

Alien crosstalk (AXT)
An unwanted signal coupling from one balanced twisted-pair component, channel, or permanent link to another twisted-pair component, channel, or permanent link

American National Standards Institute (ANSI)
A private, non-profit membership organization focused on meeting the standards and conformity assessment requirements of its diverse constituency; provides a neutral forum for the development of consensus agreements on issues relevant to voluntary standardization; the U.S. representative to the ISO, and through the U.S. National Committee, to the IEC

American wire gauge (AWG)
A system used to specify wire size; the greater the wire diameter is, the smaller the value (for example, 24 AWG is 0.51 millimeter (0.02 in.) in diameter)

Ampere
Unit of electric current; one ampere is equal to the current produced by one volt acting through a resistance of one ohm

Analog signal
A signal that uses continuous physical variables such as voltage amplitude or frequency variations to transmit information

Anchor

In a premises environment, a device inserted into a prepared hole and set, where a screw or bolt needs to be placed

Approved ground

A ground that has been approved for use by the authority having jurisdiction

Aramid

A liquid crystal polymer material with exceptional tensile strength and coefficient of thermal expansion near that of glass; widely used as a strength member in optical fiber cables; also referred to as aramid yarn

Architectural assemblies

Walls, partitions, or other barriers that are not load-bearing

Architectural structures

Walls, floors, floor/ceilings, and roof/ceilings that are load-bearing

Attenuation

Signal loss; the decrease in magnitude of transmission signal strength between points, expressed as the ratio of output to input; measured in decibels (dB), usually at a specific frequency for copper or wavelength for optical fiber; the signal strength may be power or voltage

Attenuation-to-crosstalk ratio (ACR)

The difference between attenuation and crosstalk measured in decibels (dB) at a given frequency; this difference is critical to ensure that the signal sent down the twisted-pair cable is stronger at the receiving end of the cable than any interference signals (crosstalk) from other cable pairs; the ratio of the power of the signal received (attenuated by the media) over the power of the near-end crosstalk (NEXT) from the local transmitter

Attenuation-to-crosstalk ratio, far-end (ACRF)

Pair-to-pair far-end crosstalk (FEXT) loss which is the undesired signal coupling between adjacent pairs at the far end (the opposite end of the transmit end) of cabling or a component

Authority having jurisdiction (AHJ)

The building official, electrical inspector, fire marshal, or other individual(s) or entity responsible for interpretation and enforcement of local building and electrical codes; this authority is usually represented by an electrical inspector, building inspector, or fire marshal

B

Backboard

A panel (for example, wood or metal) used for mounting connecting hardware and equipment; used to hold and support cross-connect wires routed between terminal blocks

Backbone

A facility (for example, pathway, cable or conductors) between any of the following spaces: telecommunications rooms, telecommunications enclosures, common telecommunications rooms, floor serving terminals, entrance facilities, equipment rooms, and common equipment rooms

Backbone bonding conductor (BBC)

The conductor that interconnects the riser elements of the telecommunications grounding infrastructure; utilized in larger multi-story commercial buildings with multiple telecommunications backbone risers to equalize potential between them (previously known as the grounding equalizer, or GE)

Backbone cabling

Consists of cable and connecting hardware that provide interconnections between telecommunications rooms, equipment rooms, and entrance facilities

Backbone pathway

The portion of the pathway system that permits the placing of backbone cables between the entrance location and all cross-connect points within a building and between buildings

Balun

A balanced-to-unbalanced circuit coupling device, used to convert from unbalanced to balanced transmission, and which provides impedance matching for connecting twisted-pair to coaxial cable

Bandwidth

A range of frequencies, usually the difference between the upper and lower limits of the range, expressed in hertz (Hz); used to denote the potential capacity of the medium, device, or system; in copper and optical fiber cabling, the bandwidth decreases with increasing length; the width of a communications channel; in analog systems it is measured in megahertz, or MHz (cycles per second) while in digital systems, it is measured in megabits per second (Mb/s)

Barrier

A partition installed in a raceway or cable tray that provides complete separation of the adjacent compartment

Bay

A regular repeated spatial architectural element defined by beams or ribs and their support

BNC

A bayonet locking connector used with 10Base-2 thin coaxial cable segments; these connectors, used throughout the cable length, attach to T-connectors, which in turn connect to network devices; also used in CCTV and other video applications, and used in most data coaxial applications

Beam clamp

Device made to attach to a beam or other building structure above the ceiling to hold cable supports or equipment

Bearing plate

A steel plate placed under one end of a beam, column, or truss at a support point for load distribution

Bearing wall

A wall supporting a load other than its own weight

Bel

A measure of analog signal strength; named in honor of telephone pioneer Alexander Graham Bell (the "B" in the unit of measure "dB" is capitalized because of this)

Bend radius

Maximum radius that a cable can be bent without causing physical or electrical damage or adverse transmission performance

BICSI ®

A telecommunications association, formerly known as Building Industry Consulting Service International

Bidders' conference

A meeting conducted by the issuer of a request for quote to answer questions from respondents (bidders)

Bill of materials

A list of the quantity and specific types of materials to be utilized on a project; this list should also consider exempt materials (screws, bolts, etc.)

Binary

Indicates a state or condition, such as current flow or no current, on or off; a logical one or a logical zero

Binary digit (bit)

Zeros and ones used to represent data processed by digital computing devices

Binder group

A group of wire pairs found in a large cable; groups can be distinguished from one another through the use of colored ribbons; standard cable construction methods place 25 pairs in each binder group within a cable

Bit

A binary digit, the smallest element of information in binary systems; it is either a logical one (1) or zero (0), also known as "an on or an off bit" of binary data; the smallest unit of information that a computer can process

Bit error rate

The ratio of incorrectly transmitted bits to total transmitted bits; a primary specification for all transmission systems, it is usually expressed as a power of 10; the number of errors made in a digital transmission as compared to complete accuracy

Bit rate

Transmission of a binary signal measured in bits per second

Bit stream

A series of binary digits (zeros and ones) representing the message being transferred between devices

Bonding (Bonded)

The permanent joining of metallic parts to form an electrically conductive path that will assure electrical continuity, the capacity to safely conduct any current likely to be imposed, and the ability to limit differences in potentials; connected to establish electrical continuity and conductivity

Bonding conductor

A conductor used specifically for the purpose of bonding; not intended to carry electrical currents under normal conditions, but must be capable of carrying fault currents that may result from transient voltages, power faults, or lightning strikes

Bridged tap

Multiple appearances of the same cable pair at several distribution points

Bridging clip

A metal clip utilized to couple cable conductors on a 66-series connecting block and to provide a point of physical disconnection

Broadband

A general term for transmission of signals that have wide bandwidth or multiple modulated channels

Building core

A three-dimensional space, permeating one or more floors of the building and used for the extension and distribution of utility services (for example, elevators, washrooms, stairwells, mechanical and electrical systems, and telecommunications) throughout the building

Building grounding electrode system

A network of grounded building components (for example, metal underground water piping, metal building frame, concrete-encased electrode, ground rods, etc.)

Building module

The standard selected as the dimensional coordination for the design of the building; usually a multiple of 100 millimeters (4 in), as the international standards have established a 100-millimeter (4 in) basic module; produces modular coordination to all building materials, products, and utilization of the floor space

Byte

A data unit made up of eight bits; sometimes referred to as an octet

C

Cable

An assembly of one or more insulated conductors or optical fibers within an enveloping jacket

Cable rack

The vertical or horizontal open support structure (usually made of aluminum or steel) that is attached to a ceiling or wall

Cable reel

Spool that cable is wrapped around

Cable run

A length of installed media, which may include other components along its path

Cable sheath

A covering over the optical fiber or conductor assembly that may include one or more metallic members, strength members, or jackets

Cable support system

A combination of conduits, cable trays, support hooks, hook and loop tape, and any other hardware pieces used in a cabling installation to support cables; cable support systems keep excess stress off the cables and may provide some mechanical protection to the cables being supported

Cable termination

1. Process of attaching the pairs of a cable to allow for connecting the cable to other cables or devices; examples of cable termination hardware are patch panels, connecting blocks, termination blocks (66M-, 110-, or BIX-type), and modular jacks

2. The connection of the wire or optical fiber to a device, such as equipment, panels, or a wall outlet

Cable tray, cable runway

A support mechanism used to route and support telecommunications cable or power cable; typically equipped with sides that allow cables to be placed within the sides over their entire length; a cable support system that may be used for above ceiling or in access floor pathway applications

Cable trough

A raceway consisting of metal or plastic raceway and fittings, formed and constructed so that insulated conductors and cables may be readily installed or removed without injury either to conductors or their coverings

Cabling

A combination of all copper and optical fiber telecommunications cables, equipment/patch cords, and connecting hardware

Cabling system

A specific system of telecommunications cables, equipment/patch cords, connecting hardware, and other components that is supplied as a single entity

Campus

The buildings and grounds having legal contiguous interconnection

Capacitance

The tendency of an electronic component to store electrical energy; pairs of wire in a cable tend to act as a capacitor; the charge on one of two conductors of a capacitor divided by the potential difference between them (measured in farads)

Category

A unique number classification (for example, category 5e, category 6, etc.) which defines the transmission performance parameters of twisted-pair cabling

Category 3

100-ohm twisted-pair copper cabling operating at frequencies up to 16 MHz; generally regarded as suitable only for networks operating up to 16 Mb/s using active equipment; primary usage is for backbone cabling to support voice (but not VoIP)

Category 4

100-ohm twisted-pair copper cabling operating at frequencies up to 20 MHz; performance requirements are more severe than those for category 3; developed to support communications at 16 to 20 Mb/s over runs up to 100 meters; now considered obsolete

Category 5

100-ohm twisted-pair copper cabling operating at frequencies up to 100 MHz; performance requirements are more severe than those for category 4; designed to support applications up to 100 Mb/s; now considered obsolete

Category 5e

100-ohm twisted-pair copper cabling operating at frequencies up to 100 MHz; category 5e is an enhanced version of Category 5 with more strict specifications and testing parameters

Category 6

100-ohm twisted-pair copper cable that meets or exceeds specifications in TIA-568.2-D, *Balanced Twisted-Pair Telecommunications Cabling and Components Standard*, *Generic Telecommunications Cabling for Customer Premises*, and *ISO/IEC 11801, Generic Cabling for Customer Premises*, at frequencies up to 250 MHz; performance requirements are more severe than those for category 5e; designed with a significant improvement in bandwidth, for support of Gigabit Ethernet; can support 10 Gb/s to a limited distance up to 55 meters (165 ft)

Category 6e

An enhanced or "extended" bandwidth version of category 6 cable; there is no category 6e standard; it falls under the same installation and testing parameters and specifications as category 6 cable

Category 6A (or Augmented Category 6)

100-ohm twisted-pair copper cable that meets or exceeds specifications in TIA-568.2-D, *Balanced Twisted-Pair Telecommunications Cabling and Components Standard* at frequencies up to 500 MHz; can support 10 Gb/s applications (especially 10GBase-T) up to a maximum distance of 100 meters; performance requirements are more severe than those for category 6; designed to meet or exceed the requirements of 10 Gb/s Ethernet (10GBase-T)

Category 7

100-ohm shielded twisted-pair copper cabling that meets or exceeds specifications in ISO/IEC 11801, *Generic Cabling for Customer Premises*, at frequencies up to 600 MHz

Category 8

100-ohm shielded twisted pair copper cabling supporting 25GBase-T and 40GBase-T applications; the standard defines a permanent length limit of 24 meters (78 ft) and a maximum channel length of 30 meters (100 ft) when supporting 25 Gb/s and 40 Gb/s speeds; a full channel configuration can be supported at 10 Gb/s and lower speeds; category 8 cabling became part of the ANSI/TIA 568 standard in 2016 at frequencies up to 2,000 MHz

Ceiling distribution system

A distribution system that utilizes the space between a suspended or false ceiling and the structural surface above

Cellular floor

A floor distribution system in which cables pass through floor cells constructed of steel or concrete to provide a ready-made raceway for distribution of power and telecommunications cables

Cellular floor raceway

An assembly of hollow, longitudinal units constituting part of a floor and systematically placed for the distribution of cables

Cementitious

Cement-like materials made from a dry powder premixed or mixed with water; more adaptable to large openings than putty or caulk for firestopping

Centralized cabling

An optical fiber cabling configuration from the work area to a centralized cross-connect using pull-through cables; an interconnect or splice in the telecommunications room

Channel

The end-to-end transmission between two points to which application-specific equipment is connected, including the patch cords at the device location and at the telecommunications room; TIA standards define a channel as up to 90 meters (295 ft) of horizontal cable with connectors at the work area and telecommunications room, plus up to 10 meters (33 ft) of patch cords and equipment cords for a total of 100 meters

Channel test configuration

A test configuration used by data telecommunications systems system designers and users to verify the performance of the overall channel

Characteristic impedance

An impedance of a circuit that, when connected to the output terminals of a uniform transmission line of arbitrary length, causes the line to appear infinitely long; an impedance value calculated by applying a smoothing function (typically at least squares curve fit) to measured input impedance

Cladding

The transparent outer concentric glass layer that surrounds the optical fiber core and has a lower index of refraction than the core; provides total internal reflection and protects against scattering from contaminants at the core surface

Cleave

The process of breaking an optical fiber by a controlled fracture of the glass to obtain an optical fiber end that is flat, smooth, and perpendicular to the optical fiber axis

Coaxial cable

A cable consisting of a central metallic conductor surrounded by a layer of insulating material, which is covered by a metallic mesh or solid metallic sleeve and an outer non-conducting jacket

Code

A systematic collection of regulations and rules intended to ensure safety during installation and use of materials, components, fixtures, systems, premises, and related subjects; typically invoked and enforced through government regulation

Common equipment room (CER)

An enclosed space used for telecommunications equipment and backbone interconnections for more than one tenant in a building or campus

Common telecommunications room (CTR)

An enclosed space used for backbone interconnections for more than one tenant in a building, and which may also house equipment

Commercial building

A building, or portion thereof, that is intended for office use

Common carrier

A provider to the general public of telecommunications transmission service facilities

Completion bond
A bond that ensures a contractor will complete a project to the specifications of its request for quote within a specified time

Conductance (G)
The ability of an electrical circuit or component to pass (conduct) current; can be a source of electromagnetic interference (EMI) in a transmission line

Conduit
A rigid or flexible metallic or non-metallic raceway of circular cross-section through which cables can be pulled

Conduit run
Multiple sections of conduit connected together

Conduit system
Any combination of ducts, conduits, maintenance holes, handholes, and vaults joined to form an integrated whole

Connecting hardware
A device, or combination of devices, used to connect two cables or cable elements

Connector
A mechanical device used to provide a means for aligning, attaching, and achieving continuity between conductors or optical fibers

Consolidation point
A location where horizontal cables extending from a telecommunications room are interconnected to horizontal cables extending into modular furniture pathways

Continuity
Uninterrupted connection; the state of being continuous

Contractor
A person or company contracted to perform a specific task

Cross-connect
A facility enabling the termination of cable elements to be connected to other cable elements

Cross-connection
A connection scheme between cabling runs, subsystems, and equipment using patch cords or jumpers that attach to connecting hardware on each end

Crosstalk
Any unwanted reception of signals induced on a communication line from another communication line or from an outside source

Customer premises
Building(s), grounds, and appurtenances (belongings) under the control of the customer

Customer premises equipment
Equipment residing on customer sites (for example, PBX systems, key systems, telephone sets, modems, etc.)

D

Daisy-chained
The practice of wiring devices in series

Data
Electronically encoded information

Data network
An interconnected system of computers, peripherals, and software over which commands, files, and messages are sent and received

Data transfer rate
The rate, in bits per second, at which information is transferred between network devices over a

communications channel; sometimes referred to as throughput or operating speed; the speed at which data is transferred from one device to another

Decibel (dB)
A logarithmic unit used for expressing the loss or gain of signal strength; one dB is the amount by which the pressure of a pure sine wave of sound must be varied in order for the change to be detected by the average human ear; a unit of measure used to express a ratio between two voltages or powers

Delay
The amount of time needed for a signal to travel over a twisted pair; measured in nanoseconds

Delay skew
The disparity in the propagation delay between any two pairs within the same cable; the difference between the amount of delay for the fastest (shortest pair lengths) and slowest (longest pair lengths) between pairs within the same cable

Demarcation point
A point where the operational control or ownership changes

Designation strips
Labels placed on terminal blocks and used for the identification of twisted pairs, fibers, or circuits

Digital
A nominally discontinuous signal that changes from one state to another in a limited number of discrete steps; information transmitted or stored as a series of ones and zeros indicating an on or an off state

Direct-buried cable
A telecommunications cable designed to be installed under the surface of the earth, in direct contact with the soil

Dispersion
The broadening of light pulses along the length of a fiber

Drain wire
As part of a cable (shielded, foiled, or screened), a non-insulated conductor placed in electrical contact with the cable shield, which may be used to bond the cable shield to ground

Draw
Payment method for a telecommunications project in which the contractor receives an initial payment on completion of the contract and makes periodic draws during the term of the project

D-ring
Wire management ring made of metal or plastic, shaped like the letter D, used for routing and supporting cables and/or cross-connection wires on a backboard

Duct
1. A single enclosed raceway for conductors or cables; see also *conduit, raceway*
2. A single enclosed raceway for wires or cables usually used in soil or concrete
3. An enclosure in which air is moved; generally part of the HVAC system of a building

E

Effectively grounded
Intentionally connected to earth through a ground connection or connections of sufficiently low impedance and having sufficient current-carrying capacity to prevent the buildup of voltages that may result in undue hazard to connected equipment or to persons

Electrical room
A floor-serving facility for housing electrical equipment, panelboards, and controls

Electrical service equipment

That portion of the electrical power installation that is the service enclosure or its equivalent, up to and including the point at which the supply authority makes connection

Electromagnetic interference (EMI)

Radiated or conducted electromagnetic energy that has an undesirable effect on electronic equipment or signal transmissions

Electronic Industries Alliance (EIA)

The alliance was organized along specific electronic product and market lines, and, as a standards association, developed and published industry guidelines; the EIA ceased operations in 2011

Encoder

A device that converts data into code

Entrance facility (EF)

An entrance to a building for both public and private network service cables (including antennae), including the entrance point at the building wall and continuing to the entrance room or space; consists of cables, connecting hardware, protection devices, and other equipment needed to connect cables entering from outdoors to cables which are suitable and approved for use indoors

Entrance point

The point of emergence for telecommunications cabling through an exterior wall, through a floor, or from a conduit

Entrance room or space

A space in which the joining of inter- or intra-building telecommunications backbone facilities takes place

Equipment cable, cord

A cable or cable assembly used to connect telecommunications equipment to horizontal or backbone cabling

Equipment grounding conductor

The conductor used to connect the non–current-carrying metal parts of equipment raceways and other enclosures to the system grounded conductor, the grounding electrode conductor, or both at the service equipment

Equipment room (ER)

A centralized space for telecommunications equipment that serves the occupants of a building; equipment housed there is considered distinct from a telecommunications room because of its nature or complexity

Exothermic

A chemical change that is accomplished through the release of heat

Exothermic weld

A method of permanently bonding two metals together by a controlled heat reaction, resulting in a molecular bond

Expansion joint

A joint between adjoining surfaces (for example, concrete, conduit) arranged to permit expansion and contraction with changes in temperature

F

F connector

The connector used with coaxial cable to attach drop cables to CATV taps and other devices

Far-end crosstalk (FEXT)

A type of crosstalk, expressed in decibels (dB), that occurs when signals on one twisted pair are coupled to another twisted pair within the same cable sheath as they arrive at the far end of a multi-pair cable system

Ferrule

The alignment sleeve portion of an optical fiber connector used to protect and align the stripped fiber

Fiber

Thin filament of glass or plastic that conducts a light signal; made of dielectric material consisting of core and cladding, which allows total internal reflection of light for propagation

Fiber optic cable

A cable containing one or more optical fibers; other components of the cable usually include the sheath, strength members, and buffer

Fireproof

A property of a material such as masonry, block, brick, concrete, and gypsum board that does not support combustion even under accelerated conditions

Fire-rated caulks

Materials which are easily applied and have a short installation time and a limited shelf life; set up as they cure, and re-entry requires a patch

Fire-rated door

A door assembled of various materials and types of construction used in wall openings to retard the passage of fire; these doors are rated in hours or fractions of hours

Fire resistance

A material or structure that can withstand fire and the passage of flame for some known period of time

Fire-resistance rating

The time in hours, or fraction of an hour, that a material or assembly of materials will withstand the passage of flame and the transmission of heat when exposed to fire under specified conditions of test and performance criteria

Fire retardant

Any substance added to delay the start of ignition or fire or slow the spread of the flame of any material

Fire shield

A material, device, or assembly of parts used to prevent propagation of flames, smoke, water, or gases from one cable system or pathway to an adjacent cable system or pathway (for example, between two parallel cable trays or between layers in vertically stacked trays)

Firestop

Products designed to restore the integrity of fire-rated walls and floors after they have been penetrated

Firestop system

A specific construction consisting of the material(s) (firestop penetration seals) that fill the opening in the wall or floor assembly and any items that penetrate the wall or floor, such as cables, cable trays, conduit, ducts, pipes, and any termination devices, such as electrical outlet boxes, along with their means of support

Firestopping

The process of installing listed, fire-rated materials into penetrations in fire-rated barriers to reestablish the fire-resistance rating of the barrier; a system used to prevent the spread of fire, smoke, toxic fumes or water from passing through fire-rated walls, floors, and floor/ceiling assemblies by reestablishing the fire-resistance rating of the barrier

Firewall

A wall that helps prevent fire spreading from one fire zone or area to another, and that runs from structural floor to structural ceiling

Fire zone

A contained area completely enclosed by fire-resistant rated walls, floors, and ceilings

Floor slab

1. The upper part of a reinforced concrete floor that sits on beams
2. A concrete mat poured on subgrade serving as a floor rather than as a structural member

Foam

Firestopping systems which are silicone-based and require precision mixing of two components; effective on large openings

Foil shield

A thin metallic tape wrapped around the cable core and acting as a shield against electromagnetic interference

Foiled twisted-pair (FTP) cable

A cable medium with four pairs of twisted insulated copper cable, surrounded by an overall aluminum foil and bound in a single plastic sheath; the twist screens the cable for electromagnetic interference up to 30 MHz while the foil screens for frequencies above 10 MHz; see also *screened cable*

Frequency

The measure of the number of cycles (waves) per second, expressed in hertz (Hz)

Furniture cluster

A contiguous group of work areas, typically including space divisions, work surfaces, storage, and seating

Fuse

An overcurrent protective device with a circuit-opening fusible element that is severed (open) when heated by the passage of an overcurrent; fuses are normally one-time devices: once they are open, they are not reusable

Fusion splice

A permanent joint accomplished by applying localized heat sufficient to fuse or melt the ends of two optical fibers together, forming a continuous single fiber

G

Graded index fiber

An optical fiber design in which the refractive index of the core is lower toward the outside of the fiber core and increases gradually in layers toward the center of the core; the refractive index reflects the rays inward and allows them to travel faster in the lower index of refraction region; the purpose is to reduce modal dispersion and thereby increase fiber bandwidth

Graded index profile

The refractive index profile of an optical fiber; the index of refraction decreases continuously from the inside toward the outside of the core

Ground

A conducting connection, whether intentional or accidental, between an electrical circuit (for example, telecommunications) or equipment and the earth or to some conducting body that serves in place of earth

Ground electrode

A conductor, usually a rod, pipe, or plate (or group of such conductors), in direct contact with the earth, providing a connection point to the earth

Grounded

Connected to earth or to some conducting body that serves in place of the earth

Grounded conductor

A system or circuit conductor that is intentionally grounded

Grounding

Connected (connecting) to ground or to a conductive body that extends the ground connection

Grounding conductor

A conductor used to connect electrical equipment to the grounding electrode to the building's main grounding busbar

Grounding electrode

A conductor (for example, a metal water pipe, building steel, or group of conductors) in direct contact with the earth for the purpose of providing a low-resistance connection to the earth

Grounding electrode conductor (GEC)

The conductor used to connect the grounding electrode to either the equipment grounding conductor and/or to the grounded conductor (neutral) of the circuit at the service equipment, or at the source of a separately derived system

Grounding electrode system

One or more grounding electrodes bonded to form a single reliable ground for a building, tower, or other similar structure

Grounding system

A system of hardware and wiring that provides an electrical path from a specified location to an earth ground point

Grounding terminal

A suitable bar, bus, terminal strip, or binding post terminal where grounding and bonding conductors can be connected

Header duct (trench duct, feeder duct)

A raceway of rectangular cross-section placed within the floor to tie distribution duct(s) or cell(s) to the telecommunications room

Hertz (Hz)

A unit of frequency equal to one cycle per second

High-rise building

A multi-story building (at least three stories) of structural steel or reinforced concrete construction

Horizontal cable

Cable that runs from the telecommunications room to a device location; run horizontally in floors or ceilings, and does not penetrate floors

Horizontal cabling

The portion of the telecommunications cabling system that extends from the work area telecommunications outlet to the horizontal cross-connect in the telecommunications room

Horizontal cross-connect

A group of connectors, such as patch panel or punch-down block, that allows equipment and backbone cabling to be cross-connected with patch cords or jumpers; *floor distributor* is the international term for horizontal cross-connect

Horizontal pathways

The supporting structures for concealing and/or supporting the cables placed between telecommunications rooms and work areas

Impedance

A unit of measure expressed in ohms; the total opposition (resistance, capacitance, and inductance) a circuit, cable, or component offers to the flow of alternating current (AC); the total of series resistance (R), series inductive reactance (X_L), and shunt capacitance (X_C) of a twisted pair

Impedance discontinuity

An abrupt change in a cable's characteristic impedance; can be caused by faulty connections, mismatched cable types, and untwisted sections on twisted-pair cables

Inductance

The opposition to change in current flow in an alternating current circuit

Infrastructure (telecommunications)

A collection of those telecommunications components, excluding equipment, that together provide the basic support for the distribution of all information within a building or campus

Innerduct

A non-metallic pathway placed within a larger pathway; also known as subduct

Input impedance

The ratio of the voltage at the sending end of the line to the current in the line at the sending end

Insert

An opening into the distribution duct or cell from which wires or cables emerge

Insert, afterset

An insert installed after the installation of the concrete floor slab or other flooring material

Insert, pre-set

An insert installed prior to the installation of the concrete floor stab or other flooring material

Insertion loss

1. The loss resulting from the insertion of a device in a transmission line; the weakening of a signal as it travels down the length of a cable; a measure of how much a signal is reduced in amplitude or strength as it is carried over a twisted pair

2. In an optical fiber system, the loss of optical power caused by inserting a component, such as a connector, coupler, or splice into a previously continuous optical path

Insulation

The dielectric material that physically separates wires and prevents conduction between them

Insulation displacement connector (IDC)

A type of wire terminating connection in which the insulation jacket is cut along the sides by the connector blades where the wire is inserted

Insulation piercing connector (IPC)

A type of wire terminating connection in which the insulation and stranded copper conductor are pierced by a sharp blade through the conductor insulation and into the stranded conductor

Interbuilding backbone

A backbone network providing communications between more than one building

Interbuilding backbone cable

Cable that runs between buildings in a campus environment

Interconnect

A location where interconnections are made

Interconnection

A connection scheme that provides for the connection of a cable to another cable or to an equipment cable

Interface

A physical point of demarcation between two devices or systems where electrical signals, connectors, and timing are defined; the place where the customer-provided equipment can be physically disconnected from the carrier service for testing

Intermediate cross-connect (IC)

The connection point between a backbone cable that extends from the main cross-connect (campus distributor) (first-level backbone) and the backbone cable from the horizontal cross-connect (floor distributor) (second-level backbone); *building distributor* is the international term for intermediate cross-connect

Intra-building backbone

A backbone network providing communications within a building

Intra-building backbone cable

Cable that runs between telecommunications rooms inside a building; can be vertical or horizontal in physical orientation, but is a backbone cable because it serves telecommunications rooms

Intumescent firestop

A firestopping material that expands under the influence of heat

Intumescent sheets

A firestopping material; those with sheet metal backing are used mostly for large penetrations in walls and floors; those with non-sheet metal backings may be used with caulk or putty to fabricate a honeycomb-like partitioned opening for cable, conduit, metal, or non-metallic pipe

Intumescent wrap sheets

Used for firestopping plastic piping, insulated metal piping, cable, cable bundles, non-metallic conduit, exposed innerduct, or any other material that may burn away in a fire and leave a significant void

J

J-hook

An open top cable support for horizontal cables that is shaped like a J, attached to some building structures; horizontal cables are laid in the opening formed by the J to provide support for the cables

Jumper

1. An assembly of twisted pairs without connectors, used to join telecommunications circuits/ links at a cross-connect
2. A fiber optic cord with connectors installed on both ends

Jumper wire

1. A twisted pair or pairs, without a jacket
2. Provides electrical bonds between the drain wires and ground

L

Ladder rack

A device similar to a cable tray but more closely resembling a single section of a ladder; constructed of metal with two sides affixed to horizontal cross members

Length test

A test that measures the length of the link/channel on the conductor pair having the shortest delay

Light amplification by stimulated emission of radiation (laser)

A device that produces coherent, highly directional light with a narrow range of wavelengths used in a transmitter to convert information from electrical to optical form

Light-emitting diode (LED)

A semiconductor diode that spontaneously emits incoherent light from the p-n junction when forward current is applied; it converts information from electrical to optical form; typically has a large spectral width; gives moderate performance at lower prices than laser diodes; commonly used with multi-mode fiber in data enterprise and industrial applications

Listed

Equipment included in a list published by an organization, acceptable to the authority having jurisdiction, that maintains periodic inspection of production of listed equipment, and whose listing states either that the equipment or material meets appropriate standards, or has been tested and found suitable for use in a specified manner

Local area network (LAN)

A geographically limited data communications system for a specific user group consisting of a group of interconnected computers sharing applications, data, and peripheral devices such as printers and CD-ROM drives intended for the local transport of data, video, and voice

Loss

Attenuation of a signal, usually measured in decibels (dB)

M

Main building ground electrode

The designated point to which all utilities in a building are connected

Main cross-connect (MC)

The cross-connect normally located in the (main) equipment room for cross-connection and interconnection of entrance cables, first-level backbone cables, and equipment cables; *campus distributor* is the international term for main cross-connect

Main distribution panel

Electrical service entrance facility

Mechanical splicing

In reference to optical fibers, permanently joining two fibers together by mechanical means to enable a continuous signal

Modal dispersion

Dispersion, or pulse spreading, resulting from the different optical path lengths (modes) in a multi-mode optical fiber

Mode

1. Loosely, a possible light path followed by light rays, as in multi-mode or single-mode
2. Strictly, a distribution of electromagnetic energy that satisfies Maxwell's equations and boundary conditions in guided wave propagation, such as through a waveguide or optical fiber

Modular jack

A telecommunications connector that accepts a modular plug

Modular patch panel

An organizational termination point for LAN cabling which allows the installer to field-configure for various networking requirements including multimedia installations; also called an unloaded patch panel

Modular plug

A telecommunications connector for cable or cords intended to plug in to a modular jack

Modulation

A process whereby certain characteristics of a wave, often called the carrier, are varied or selected in accordance with a modulating function; includes amplitude, frequency, or phase, and other modulation techniques

Monolithic pour

The single, continuous pouring of a concrete floor and columns of any given floor of a building structure

Multi-mode

Transmits or emits more than one propagating mode

Multi-mode fiber (MM)

An optical fiber that can carry multiple signals (distinguished by frequency or phase) at the same time; a graded-index or step-index optical fiber that supports the propagation of more than one bound mode

Multiplex

Combining two or more signals into a single wave (the multiplex wave) from which the signals can be individually recovered

Multiplexer

A device that combines two or more signals over a single communications channel (for example, time-division multiplexing and wavelength-division multiplexing)

Multiplexing

The combining of two or more communications channels into a common, high-capacity channel from which the original signals may be individually recovered

Multi-user telecommunications outlet assembly (MUTOA)

Terminates multiple horizontal cables in a common location within a furniture cluster

N

Nanometer (nm)

A unit of measurement equal to one billionth of a meter; the most common unit of measurement for optical fiber operating wavelengths

National Electrical Code ®

A safety code written and administered by the National Fire Protection Association

National Fire Protection Association (NFPA)

Association that writes and administers the *National Electrical Code* ®

N-connector

A coaxial connector that may be used with larger (RG-8, RG-11, and 10base-5) coaxial cables; N-type connectors have a center pin that must be installed over the cable's center conductor

Near-end crosstalk (NEXT)

The undesired coupling of signal energy from a transmitting conductor pair into a receiving conductor pair nearest the point of transmission

Network interface card (NIC)

A network interface device in the form of a circuit card that is installed in an expansion slot of a computer to provide the means to physically connect stations and servers to the network communications channel; also called adapter card or network adapter card

Noise

An unwanted electrical signal on a wire that provides a random or persistent disturbance that interferes with the clarity or quality of the expected signal and alters the shape of the signal; see also *electromagnetic interference*

Nominal velocity of propagation (NVP)

The approximate speed at which a signal moves through a cable

O

Ohm (Ω)

The standard unit of electrical resistance that measures the opposition to the flow of direct current, called resistance, or opposition to the flow of alternating current, called impedance; one volt will cause one ampere of current to flow through one ohm of resistance or impedance

Open office

An area of floor space with division provided by furniture, movable partitions, or other temporary means instead of by building walls

Open office cabling

The cabling that distributes from the telecommunications room to the open office area utilizing a consolidation point or multi-user telecommunications outlet

Outlet box

A metallic or non-metallic box mounted within a wall, floor, or ceiling and used to hold telecommunications outlets/connectors or transition devices

P

Pairs in Metal Foil (PiMF)

Type of shielded cable which has a foil shield for each pair to virtually eliminate internal crosstalk and usually includes an overall foil or screen to provide additional EMI immunity

Patch cord

A length of cable with connectors on one or both ends used to join telecommunications circuits/links at the cross-connect

Patch panel

An organizational tool used in a LAN to terminate twisted-pair cables on a 110-type termination block for connection to a switch or other equipment via a patch cable

Pathway

A facility for the placement of telecommunications cable

Penetration

An opening made in a fire-rated barrier (architectural structure or assembly); there are two kinds of penetrations: a *through penetration* that extends completely through the barrier, and a *membrane penetration* that penetrates only one side of a barrier

Performance bond

A bond that ensures a contractor will use specified methods and procedures in performing a project

Permanent link

The transmission path between any two interfaces of cabling, excluding patch cords, equipment, and work area cables

Permanent link test configuration

A test link configuration used by data telecommunications systems designers and users to verify the performance of permanently installed cabling

Primary bonding busbar (PBB)

The main telecommunications grounding and bonding busbar located in the entrance facility and connected to the AC ground (previously known as the *telecommunications main grounding busbar,* or TMGB)

Plaster ring

A metal device used to adapt a four- or five-inch box to accept wiring devices and/or faceplates on a plaster or drywall wall or ceiling; available in multiple raise sizes to accommodate different thicknesses of wall coverings for the purpose of mounting a telecommunications faceplate

Plastic Insulated Conductors

Plastic conductors which ushered in the use of tip and ring color code

Plenum

A compartment or chamber to which one or more air ducts are connected and that forms part of the air distribution system; cables installed in this space require a higher fire rating

Plenum cable

A cable with flammability and smoke characteristics that meet the safety requirements allowing it to be routed in a plenum area without being enclosed in a conduit

Point of demarcation

See *demarcation point*

Poke-through device

An assembly that allows through-penetration of floor decking with telecommunication cables, or power, or both, while maintaining the fire-rating integrity of the floor

Primary bonding backbone (PBB)

Serves as the dedicated extension of the building grounding electrode system for telecommunications infrastructure; also serves as the central attachment point for telecommunications bonding backbone(s) and equipment

Power sum alien near-end crosstalk (PS ANEXT)

The measurement of alien crosstalk when the disturber link is stimulated at the same side of the link where the victim link is measured

Power sum attenuation-to-crosstalk ratio, far-end (PSACRF)

The undesired coupling of signal energy from a number of simultaneous-transmitting pairs into a receiving pair measured at the far end or the opposite end from the transmitter

Power sum attenuation-to-alien-crosstalk ratio, far-end (PS AACR-F)

The measurement of alien crosstalk ratio, corrected for the length of the link when the disturber link

is stimulated from the near end of the link and the victim link is measured from the far end of the link

Power sum near-end crosstalk (PSNEXT)

The undesired coupling of signal energy from a number of simultaneously transmitting pairs into a receiving pair; the sum of the NEXT power from all other pairs in the cable at the near end of a cable, link, or channel

Premise network

A transmission network inside a building or group of buildings that connects various types of voice and data communications devices, switching equipment, and other information management systems to each other and to outside communications networks

Pre-set insert

See *insert, pre-set*

Private branch exchange (PBX)

A device allowing private local voice (and other voice-related services) switching over a network

Protector

A device consisting of one or more protector units intended to limit abnormal surges on metallic communications circuits; includes a mounting assembly for the protector units

Protector unit

A device intended to protect against either overvoltage or overcurrent, or both; the unit may contain carbon arresters, gas tubes, solid state devices, heat coils, positive temperature coefficient devices, or a combination of these components for a specific application

Public network

A network operated by common carriers for the administration of circuits for use by the public

Punch down

The process of terminating copper cable conductors on insulation displacement connection terminals by use of a handheld tool

Putty

A popular firestop wherever frequent cable revisions are likely; most are intumescent (they expand or swell when exposed to heat)

R

Raceway

Any enclosed channel designed for holding wires, cables, or busbars

Rack

A vertical metallic frame that is equipped with threaded holes on the front of the rack (or on both the front and rear sides of the rack); it is used to mount termination hardware, electronic equipment, or a combination of both; it can be floor mounted (free standing) or wall mounted

Rack bonding busbar

A central point for attaching equipment grounding connections on a telecommunications rack; available in horizontal or vertical configurations

Radio frequency interference (RFI)

A disturbance in the reception of radio and other electromagnetic signals due to conflict with undesired signals

Rayleigh scattering

The collision of the light modes and the impurities within the core of the fiber; this accounts for a large amount of the attenuation in fiber optic cables

Refraction

The angular change in direction of a beam of light at an interface between two dissimilar media or a medium whose refractive index is a continuous function of position (graded-index medium)

Refractive index

The ratio of the velocity of light in a vacuum to the velocity of light in a given material; the ratio of the velocity of light in the core and cladding of the same fiber

Request for proposal (RFP)

A document that solicits quotes for telecommunications cabling projects or equipment and provides vendors with the information necessary to prepare a bid

Resistance

A measure of opposition a material offers to the flow of direct current

Return loss (RL)

The ratio of the transmitted signal to the reflected signal of the cabling, expressed in decibels (dB); a measurement of the amount of signal energy that is reflected back to a transmitter as it travels down a twisted pair while encountering impedance discontinuities

Ring

A means for identification of one conductor of a pair; historically associated with the wire connected to the "ring" portion of an operator's telephone plug

Riser

Term applied to vertical sections of cable, such as changing from underground or direct-buried plant to aerial plant; also applies to the space used for cable access between floors

S

Screened twisted-pair (ScTP) cable

A cable with one or more pairs of twisted copper conductors covered with an overall metallic screen or braid over the group of four pairs to provide AXT and EMI shielding

Secondary bonding busbar (SBB)

Serves as the bonding connection point for the systems and equipment within the telecommunications room or space; it is directly connected to the primary bonding busbar via the telecommunications bonding backbone (previously known as the *telecommunications grounding busbar*, or TGB)

Service equipment

The necessary equipment, usually consisting of a circuit breaker or switch and fuses, and their accessories, located near the point of entrance of supply conductors to a building or other structure, or an otherwise defined area, and intended to constitute the main control and means of cutoff of the electrical supply

Shield

Metallic layer placed around a conductor or group of conductors to prevent electrostatic or electromagnetic coupling between the enclosed wires and external fields

Shielded twisted-pair (STP) cable

Cable made up of multiple twisted copper pairs where the entire structure is covered with an overall shield or braid and an insulating sheath (cable jacket)

Single-mode optical fiber (SM)

An optical fiber that supports only one mode of light propagation with a core size between eight and 10 microns

Single-point ground terminal

A connecting point provided with PBX and key systems; the only acceptable point for connection from the equipment to the telecommunications grounding system

Slot

An opening (usually rectangular) through a wall, floor, or ceiling to allow the passage of cables and wires

Space (telecommunications)

An area used for housing the installation and termination of telecommunications equipment and cable; for example, common equipment rooms, equipment rooms, common telecommunications rooms, telecommunications rooms, work areas, and maintenance holes/handholes

Spade lug

A U-shaped metal connector that is soldered or crimped to a wire; used for connection to a terminal post

Splice

A permanent joining of conductors, generally from separate sheaths, using mechanical means

Splice closure

A metal or plastic housing with a cavity used to clamp around a cable splice to provide environmental protection

Suspended ceiling

A ceiling that creates an area or space between the ceiling material and the building structure above the material; area may or may not be an air handling space (plenum)

Switch

A communications device that utilizes switching technology to establish and terminate connections

T

Telecommunications

A branch of technology concerned with the transmission, emission, and reception of signs, signals, writing, images, and sounds; that is, information of any nature by cable, radio, optical, or other electromagnetic systems

Telecommunications bonding backbone (TBB)

A conductor that interconnects all secondary bonding busbar(s) with the primary bonding busbar

Telecommunications bonding conductor (TBC)

Interconnects the building's service equipment (power) ground to the telecommunications grounding system; shall bond the primary bonding busbar to the service equipment (power) ground; at a minimum the same size as the TBB (previously known as the *bonding conductor for telecommunications*)

Telecommunications equipment bonding conductor (TEBC)

Should be installed from each piece of equipment to the telecommunications secondary bonding busbar (SBB) or telecommunications primary bonding busbar (PBB)

Terminal block

A device that serves to terminate cable conductors

Termination

The act of connecting a cable/wire/fiber to connecting hardware

Tip

A means for identification of one conductor of a pair; historically, associated with the wire connected to the tip portion of an operator's telephone plug

Transition point

A location where flat undercarpet cable connects to round horizontal cable

Trench duct

An interior or exterior trough embedded in concrete that has removable cover plates level with the top of the surrounding surface

Trough

A pathway for the containment of cable, typically provided with a removable cover

Twisted pair

Two individually insulated conductors physically twisted together at regular intervals

Twisted-pair cable

A cable comprising two or more twisted pairs

Two-level duct

An underfloor raceway system installed with the header raceways and the distribution raceways on two different planes

U

Underfloor raceway

A pathway placed within the floor and from which wires and cables emerge to a specific floor area

UL, LLC

A U.S.-based independent testing laboratory that sets safety tests and standards for electrical equipment

Unshielded twisted-pair (UTP) cable

Four pairs of solid or stranded copper conductors, each pair twisted at a different rate to minimize crosstalk (interferance), all within the same unshielded cable jacket

Volt (V)

A unit of electromotive force or potential difference that will cause a current of one ampere to flow through a resistance of one ohm

Wire map test

A basic continuity test to examine for open or short circuits, crossed pairs, reversed wires, and split pairs

Work area

A building space where the occupants interact with telecommunications terminal equipment

Work area cable

A cable connecting the telecommunications outlet/connector to the terminal equipment

Acronyms and Abbreviations

A

A ampere

AC alternating current

ACEG alternating current equipment ground

ACR attenuation-to-crosstalk ratio

ACRF attenuation-to-crosstalk ratio, far-end

ADA Americans with Disabilities Act

AHJ authority having jurisdiction

AMES architectural, mechanical, electrical, structural

ANSI American National Standards Institute

ANSI/TIA/EIA American National Standards Institute/Telecommunications Industry Association/Electronic Industries Alliance

AWG American wire gauge

AXT alien crosstalk

B

BAS building automation system

BBC backbone bonding conductor

BICSI ® Building Industry Consulting Service International

bit binary digit

BNC Bayonet Neil-Concelman

BOCA Building Officials and Code Administrators

BOMA Building Owners Managers Association

C

CATV community antenna television (cable television)

CENELEC Comité Européen de Normalisation Electrotechnique (European Committee for Electrotechnical Standardization)

CER common equipment room

COAX coaxial cable

CO-OSP customer-owned outside plant

CP consolidation point

CSA Canadian Standards Association

CSI Construction Specifications Institute

CTR common telecommunications room

D

dB decibel

E

EF entrance facility

EIA Electronic Industries Alliance

EMC electromagnetic compatibility

EMI electromagnetic interference

EMI/RFI electromagnetic interference/radio frequency interference

EMT electrical metallic tubing

ENT electrical non-metallic tubing

ER equipment room

ESD electrostatic discharge

ETSI European Telecommunications Standards Institute

e/w equipped with

F

FEXT far-end crosstalk

FTP foil twisted-pair

G

G giga-

ga gauge

Gb/s gigabits per second

GEC grounding electrode conductor

gnd ground

H

HC horizontal cross-connect

HVAC heating, ventilation, and air conditioning

Hz hertz

I

IBC interconnecting bonding conductor

IC intermediate cross-connect

IDC insulation displacement connector

IEEE Institute of Electrical and Electronic Engineers

in inch

IPC insulation piercing connector

ISO International Organization for Standardization

K

k kilo-

km kilometer

L

LAN local area network

laser light amplification by stimulated emission of radiation

LED light-emitting diode

M

m meter

MAC move, add, or change

M mega-

Mb/s megabits per second

MC main cross-connect

m milli-

MHz megahertz

MHz/km megahertz per kilometer

mm millimeter

MM multimode

MUTOA multi-user telecommunications outlet assembly

mux multiplexer

N

NEC® *National Electrical Code®*

NECA National Electrical Contractors Association

NEMA® National Electrical Manufacturers Association ®

NESC® *National Electrical Safety Code* ®

NEXT near-end crosstalk

NFPA® National Fire Protection Association ®

NIC network interface card

nm nanometer

NVP nominal velocity of propagation

P

PBB primary bonding busbar

PBX private branch exchange

PC personal computer

PiMF pairs in metal foil

PSAACRF power sum alien attenuation-to-crosstalk ratio, far-end

PSACRF power sum attenuation-to-crosstalk ratio, far-end

PSANEXT power sum alien near-end crosstalk

PSNEXT power sum near-end crosstalk

R

RBB rack bonding busbar

RFI radio frequency interference

RFP request for proposal

RFQ request for quote

RL return loss

S

SBB secondary bonding busbar

ScTP screened twisted-pair

SM single-mode

STP shielded twisted-pair

T

TBB telecommunications bonding backbone

TCB telecommunications bonding conductor

T&C terms and conditions

TEBC telecommunications equipment bonding conductor

TIA Telecommunications Industry Association

TO telecommunications outlet

TP transition point

TR telecommunications room

TSB Telecommunications Systems Bulletin (formerly Technical Systems Bulletin)

U

USOC universal service order code

UPS uninterruptable power supply

UTP unshielded twisted-pair

W

WA work area

Wall-mounted telecom outlet box/connector (denote type, size, configuration, and descriptive information as required)

Floor-mounted telecom outlet box/connector (denote type, size, configuration, and descriptive information as required)

Ceiling-mounted telecom outlet box/connector (denote type, size, configuration, and descriptive information as required)

Equipment rack (show to scale)

Equipment cabinet (show to scale)

T Terminal cabinet, surface mounted (denote size)

Cable/Wire (denote type)

Cable to be removed

Cable slack (denote slack length)

Cap placed on cable

Cable Termination Symbols

Termination hardware, not protected
(denote terminal size)

Termination hardware, with protection
(denote terminal size)

Cross-connect hardware
(descriptive information denoted, as required)

Cross-connect hardware, not protected

Cross-connect hardware, primary protector integrated with cross-connect hardware

Cross-connect hardware, primary and secondary protector integrated with cross-connect hardware

Primary protector, not cross-connected

Primary protector, cross-connected

Grounding/Bonding Symbols

Connection to ground (conductor size noted)

Space Symbols

Handhole (denote dimensions)

Manhole (denote dimensions)

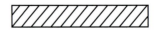

Pull box (denote dimensions)

Backboard (denote dimensions)

Telecommunications space
(denote name of space)

Splice Symbols

Splicing location

Straight splice, no changes
(singly administered)

Straight splice, with changes
(separately administered)

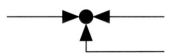

Splice with branch cable
(separately administered)

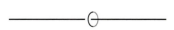

Insulated splice

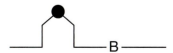

Splice, buried cable, in pedestal

Pathway Symbols

100 mm (4in)
EMT

Single conduit run, with endpoint
(denote size and type)

Single conduit home run
(denote size, type, and home location)

Single conduit run turned down
(denote size and type)

Single conduit run turned up
(denote size and type)

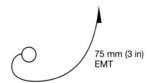
75 mm (3 in)
EMT

Conduit stub up
(denote size and type)

Conduit bank, plan view
(denote quantity, size, and type)

Conduit bank, section view
(denote quantity, size, and type)

Conduit bank, concrete encased
(denote quantity, size, and type)

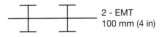

Conduit sleeve
(denote quantity, type, and size)

Backbone conduit/sleeve
(denote size and type)

Slot (denote dimensions)

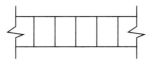

Cable tray
(denote size and type)

Utility column (interior use)
(denote type and size)

Surface-mounted raceway
(denote type and size)

Telephone pole, exterior use
(denote size, class, ownership)